# STREET ROD BUILDING SKILLS

## By John Thawley

ISBN #0-936834-32-3

**Editor**  Steve Smith
**Associate Editor**  Georgiann Smith



Published by

## STEVE SMITH AUTOSPORTS PUBLICATIONS

*P.O. Box 11631/Santa Ana, CA 92711/(714) 639-7681*

# Table of Contents

# Foreword

Our thanks to those helping with this book go in two directions — to those skillful individuals who helped supply information like Bob Lindebaum and Bob Spina; and to those countless rodders at meets all over the country who honored us by asking us how to do this or that. We, in turn, got the answers from a rodder who had been there before. Rod builders like "old man" Lindebaum are the real heroes of this activity called Rod Building. We are honored by their presence and are respectful of their skills.

**John Thawley**

# Metal Shrinking

Metal shrinking has been a stock-in-trade technique for good body men since cars started getting crunched. However, if a rod builder isn't familiar with the technique, then there may be some misunderstanding of just what is to be accomplished and how it is to be done.

Shrinking is the technique used to restore stock contours to sheet metal after that metal has been stretched out of shape. Maybe it got stretched out of shape when it came into contact with a telephone pole at 32.5 mph; or maybe it got stretched out of shape by some would-be body man who was long on the exercise of using a hammer, but a little short when it came to thinking.

To properly shrink sheet metal you'll need an acetylene torch with a small tip, a sheet metal hammer (or a hammer with a flat hammering surface — as opposed to a picking hammer) and a wet rag. The area of the sheet metal to be worked should be free of paint — and it doesn't really make much difference to the shrinking process whether you grind the paint off or strip it off with a chemical — just get the paint off the metal. Locate the largest bump on the panel to be worked. Note that we said bump — not valley or depression. The bump must go toward you as you face the panel. Heat the bump with the torch until it is cherry-red. Put the torch down quickly and *gently* hammer the bump downward. Now shove the dripping wet rag onto the area. You now know how to shrink. Take a close look at the photographs included in this section and you'll get a good idea about how a professional rod builder goes about saving a valuable panel without the use of plastic filler, replacement panels or lead.

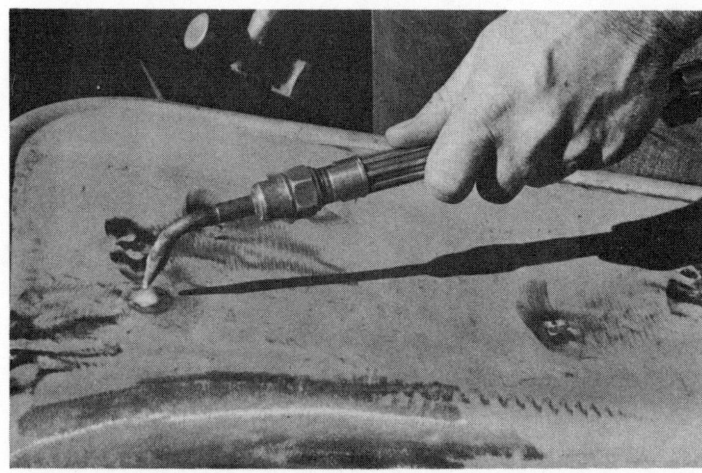

The first step in shrinking is to apply just enough heat to the largest of bumps to make the metal cherry-red. This red area should be kept to about the size of a dime.

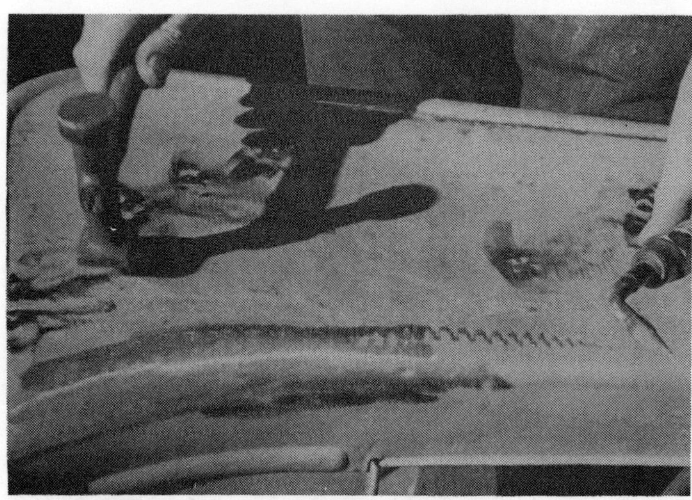

Gently hammer the bump downward. Note here that no dolly backs up the hammer and the torch remains lit and nearby for the next application of heat.

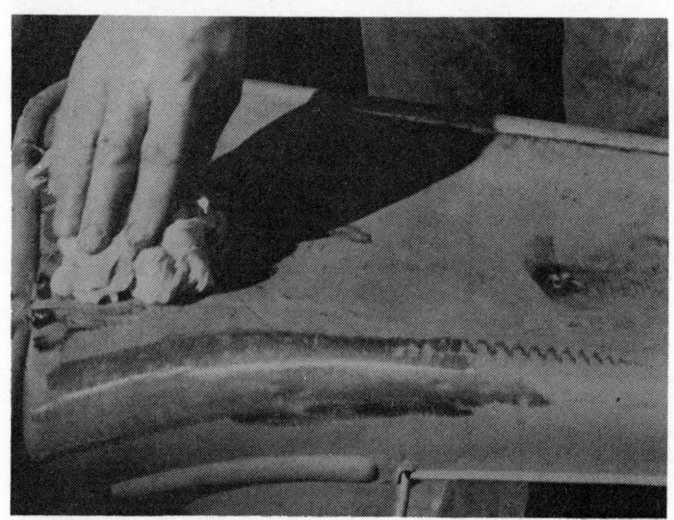

Quench the heated area with a wet rag as soon as the bump has been hammered down. Keep the rag soaking in a can of water nearby when it is not being used on the metal.

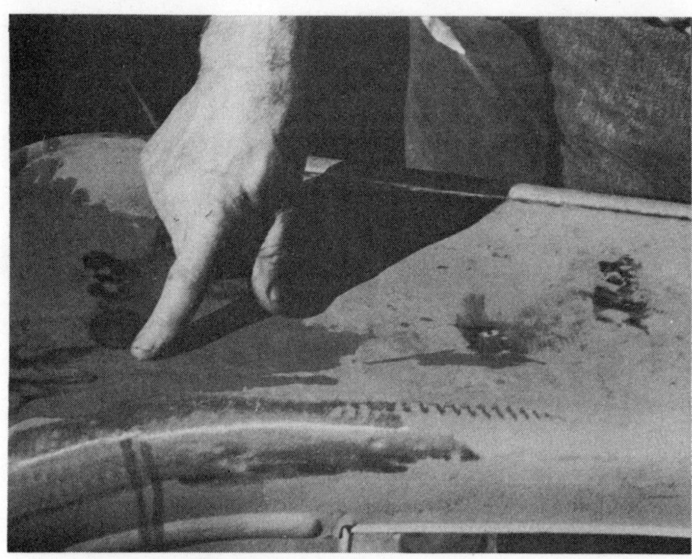

Let the metal cool and then feel where the next largest bump is. Then repeat the shrinking procedure.

After a short session of heating and hammering, the panel begins to straighten out. The dark areas indicate all of the places heat has been applied.

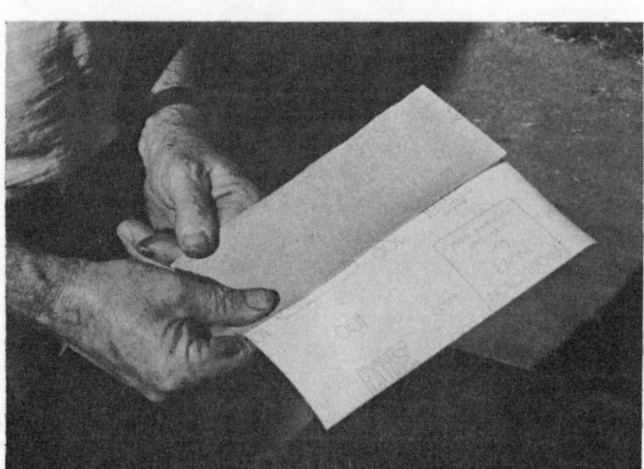

A full size sheet of 100 grit sandpaper is folded in thirds — which makes it last longer and also easier to hold.

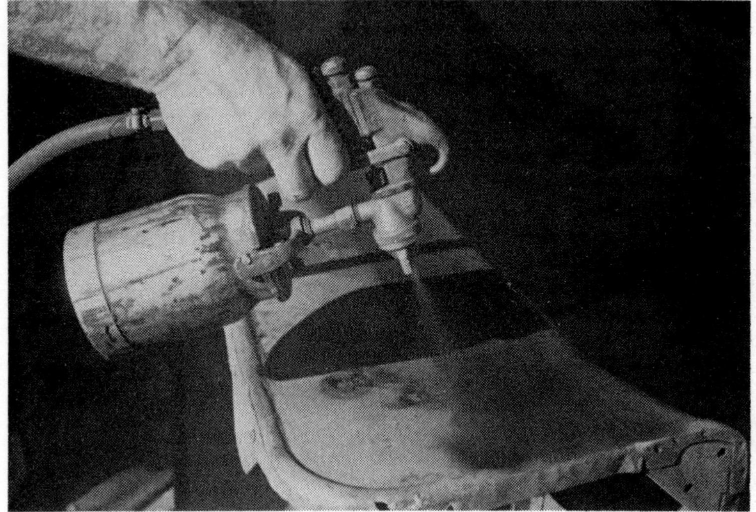

(Left) Once a panel has been worked fairly smooth with the technique of shrinking, a light coat of primer can be applied for the next step in making the panel "rod perfect."

(Below) Just a couple of swipes across the panel will reveal what might seem like an insurmountable amount of bumps.

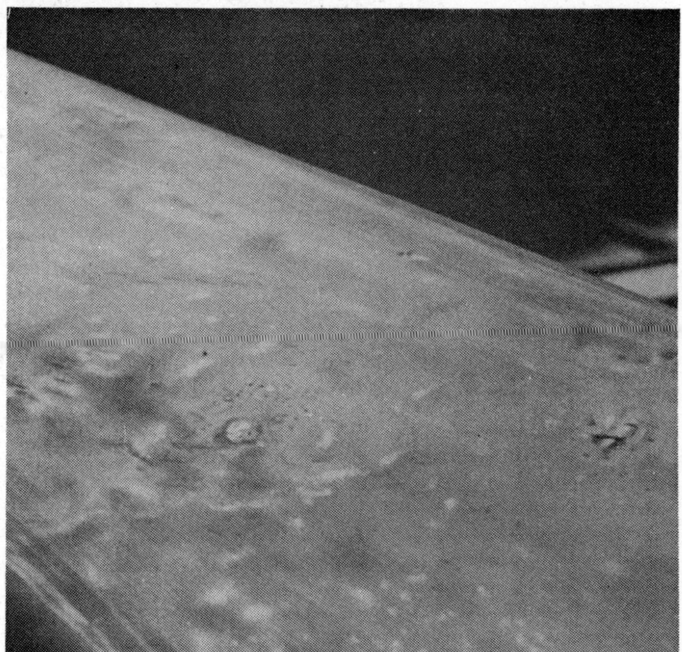

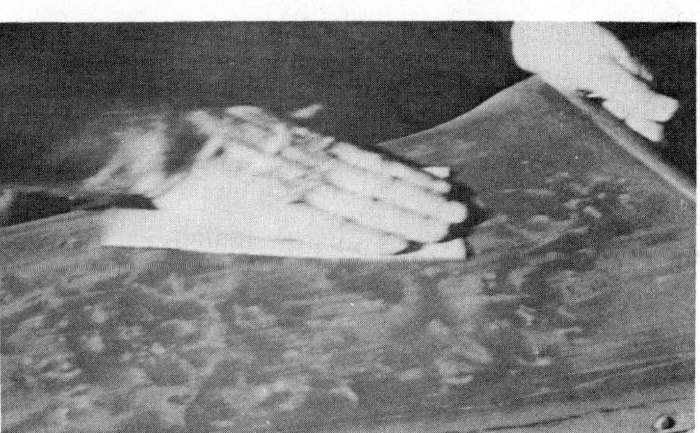

The surface of the panel is lightly scuffed simply to reveal the remaining smaller bumps. Note that they are in evidence here.

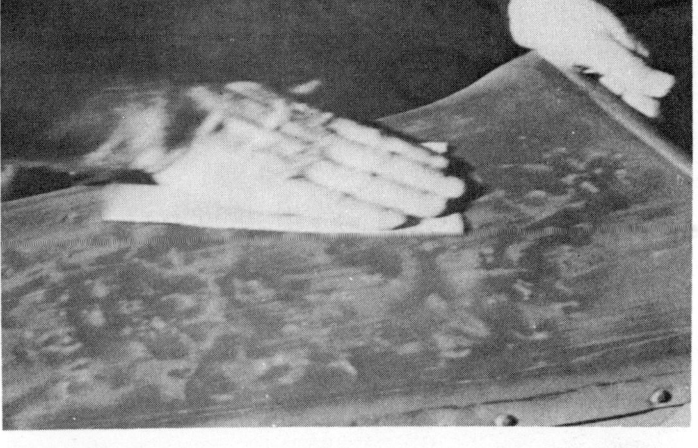

At this point, a small picking or shrinking hammer is used to tap down the smaller bumps.

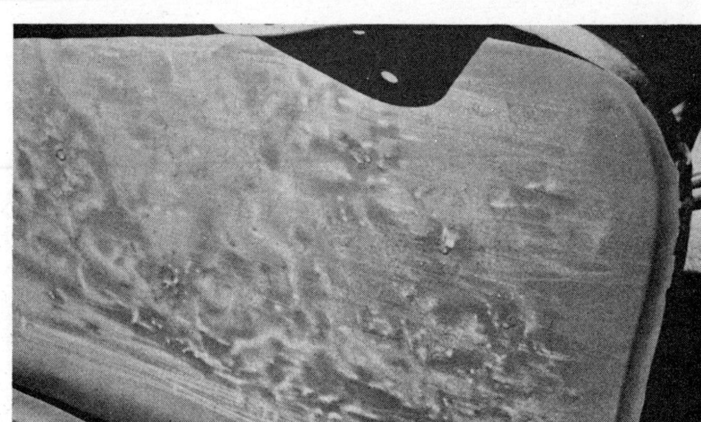

As the panel gets smoother and smoother, apply the primer and sanding procedure to the underside of the panel.

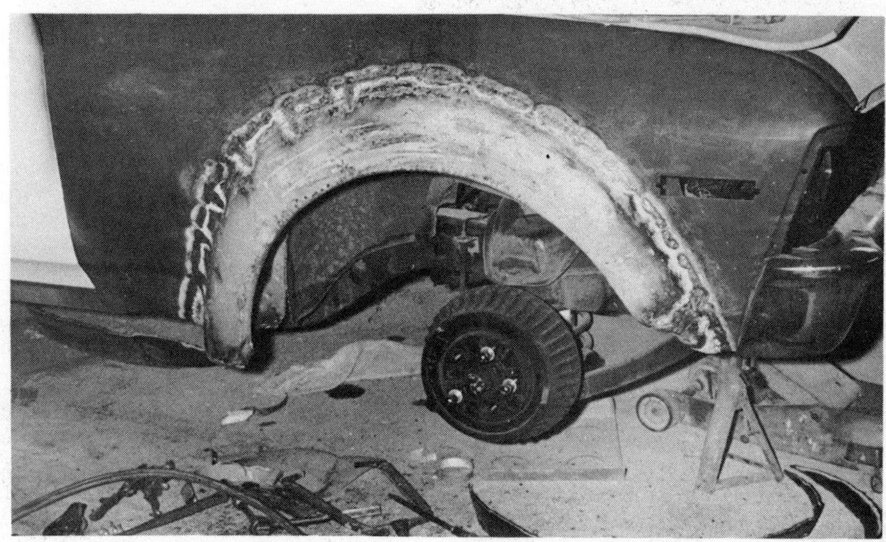

# Hammer Welding

Good hammer welding simply takes two pieces of metal and fuses them together. The result is as strong as a single piece of metal. In the case of sheet metal on a rod, hammer welding can and often does go one step further by making the joint flawless without the use of any filler. Before turning to the photos in order to get a better grasp of how an expert goes about this quite useful technique, a couple of points should made.

In the first place, if the panel being worked on has a bead or molding such as the edge of a fender, you should concentrate on making that part absolutely perfect. Make certain the two panels are lined up perfectly before welding is started. If the bead or molding is correct, the rest of the panel can be less than perfect and the flaws will not be detected because the eye will follow the bead line. Also, when joining two panels by hammer welding, be sure to leave a gap of about 1/16-inch between the two pieces. As the rod is used to fill the gap and fuse the edges, the panels pull together. This cuts down warpage at the seam and assures full penetration of the weld. When you are hammer welding one thick and one thin piece of metal together, warpage will occur at different rates within the two panels, so bring the metal back into shape before the warpage gets out of hand. Lastly, you should practice welding sheetmetal panels together until you have a feel for the metal. No book in the world can teach you perfect hammer welding.

*The first step in hammer welding is to ensure that the metal to be worked is held securely. In this case, two panels are clamped to a steel work table and as a first step the two panels are tacked together.*

To cut down on warpage, the two panels are tacked together for the full length while still being held by the clamps.

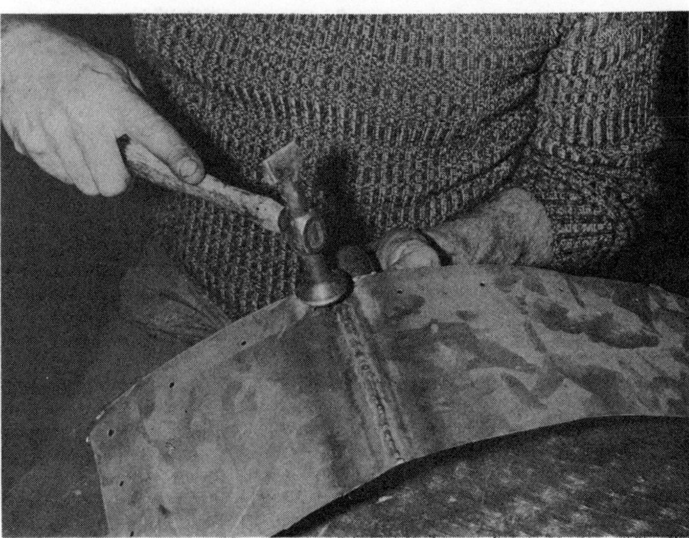

So not attempt to heat and then hammer large sections of the weld. Two inches at a time is sufficient.

As the bead is being run, the weld is worked little by little with hammer and dolly. Note that the flame is quite small. Filler rod is required when joined sheetmetal does not touch. Otherwise, fusion welding (no filler rod) is the best possible method.

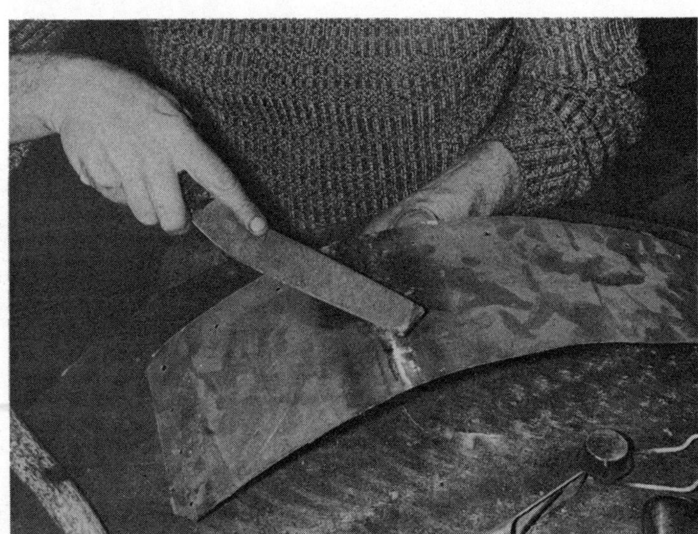

After the fusion of the two panels is complete, the surface should be worked with a body file. Excessive use of a body grinder can remove too much metal and still not reveal low portions which need to be worked.

*After the metal is worked close to perfect with file, hammer and dolly, the surface can be finished with a body grinder and jitterbug sander in preparation for paint.*

*Here's an example of what can be done with good hammer welding — sections of two fenders were joined together to make one — with not a speck of filler in sight. We had to ink in where the hammer welding took place since it cannot be detected once the job is finished.*

# Rod Building Tips I

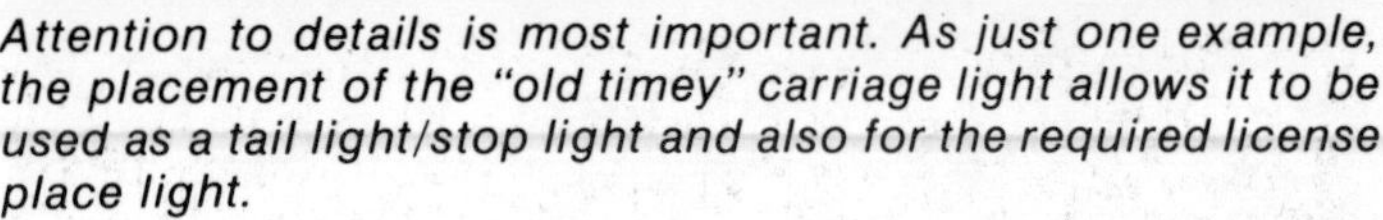

Attention to details is most important. As just one example, the placement of the "old timey" carriage light allows it to be used as a tail light/stop light and also for the required license place light.

(Top right and right) Installing directional signals on an early car can cause some head scratching if the original steering column is retained. One fairly simple and unobtrusive solution is to use a big truck directional signal which can be mounted in or under the dash. Most units come apart easily so the push buttons will read correctly if the unit is to be installed upside down.

Give some thought to a home-made luggage rack if your are short on luggage space for long distance rod runs. That substantial looking trailer hitch is another worthwile, practical idea.

Wood can often be used for an accent or highlight, and in this case it probably replaces a rusted-out panel.

Nothing fancy here — and that's the beauty of it. Wiring and plumbing are extremely neat and held to a minimum.

Don't be ashamed to seek help in the form of hardware readily available from the local auto parts store. The combination license plate light and frame is a case in point.

Here's another fine example of "simple is best" school of design. Note the pinstriping on the cowl.

*Here are a couple of ideas to keep in mind when the time comes to mount the license plate and club plaque.*

*(Right) The dash on this mid-fifties pickup was completely reshaped in order to accommodate the aftermarket engine-turned instrument panel. Woodgrain on the lower half of the dash was done with a furniture refinishing kit.*

*(Below) A polished intake manifold and unique air cleaner are attention getters on this rod. Here's still another example of a clean firewall.*

*(Below right) This '56 Ford pickup interior is sanitized to the "Nth" degree. White upholstery and plenty of chrome brighten the cab.*

*The wide expanse of a tailgate allows a great number of treatments to be pulled off. Despite the letters and aluminum insert, note the retention of stock hinges and latches.*

*The famous Niekamp roadster belonging to Jim Jacobs tells it all when it comes to the story on early fifties rod styling.*

*The old race car looks can be achieved by engine turning and the use of a multitude of gauges. Rod run plaques seem to be right at home most anywhere on a rod these days.*

*A custom dash plaque is used here to proclaim serial number, engine number and other vital statistics. Jewelry stores and trophy shops can make these at very little cost.*

*An early style steering wheel and an instrument cluster along these lines tell you that this rod was not built yesterday — even if it was.*

*Early car driving lights make excellent back up lights — for early cars.*

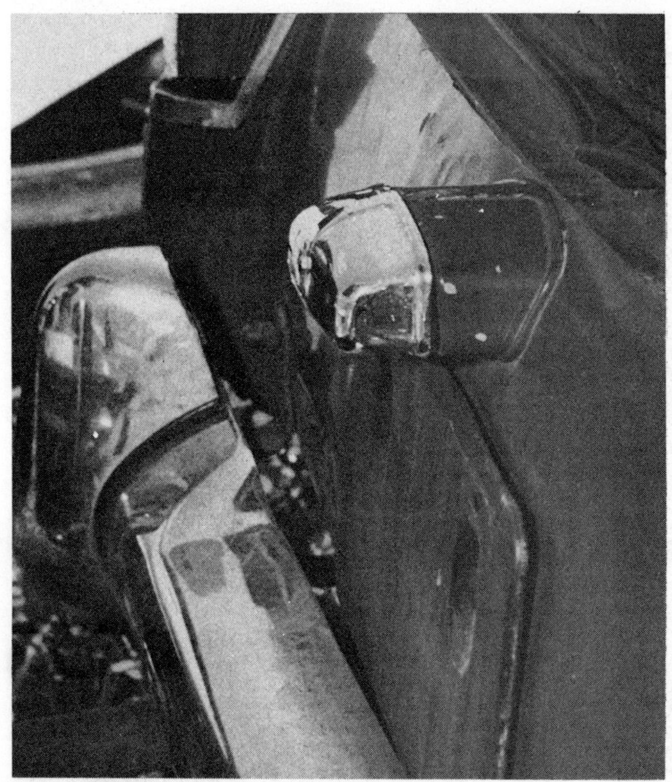

The license plate light shown here is a good example of what you can find useful in wrecking yards to put the finishing touch on a rod.

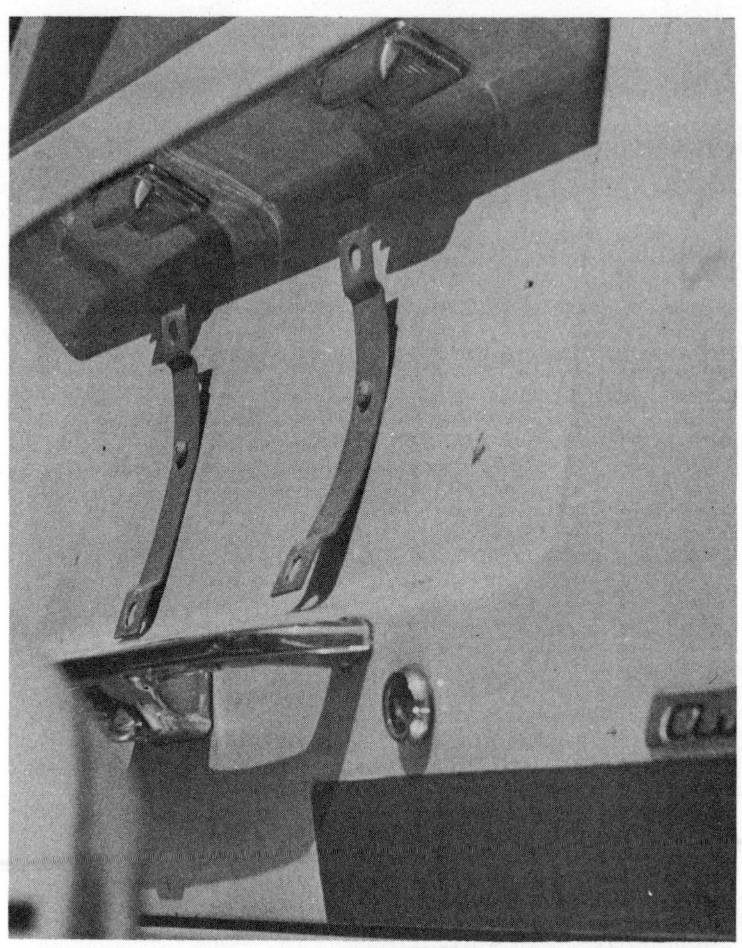

Here are two more unusual license plate lights found on a foreign station wagon. The license plate holder straps are a neat idea also.

How about this for a slick way to hide the gas cap?

*This pickup bed is actually a trailer — and as you can see a gas tank lurks inside — complete with gauge.*

*The chrome firewall seems to give this engine compartment a cluttered look — it isn't. Everything is very sanitary and a great deal of attention has been paid to detail.*

*For something a little different in carburetion this rodder adapted two sidedraft Webers to this Edelbrock four barrel manifold. Slick.*

*This rumble-seated roadster is typical of the practical, moderate cost rod that can serve as solid family transportation. Add-on trunk is perfect storage for camping trips or longer rod runs; also note the mount location for the CB antenna.*

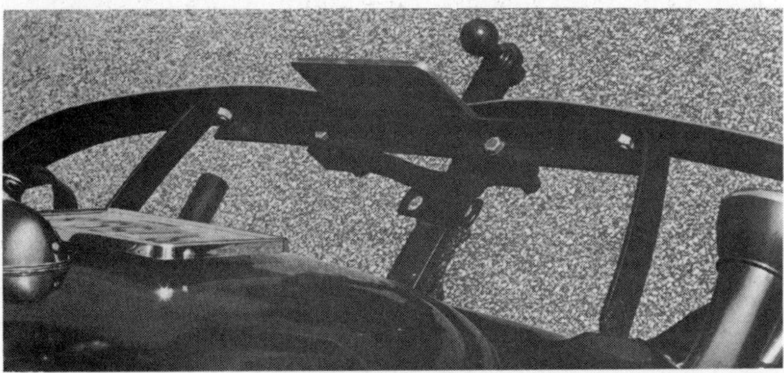

*This is a plenty slick arrangement for a trailer hitch on an early car. The brace across the backside of the bumper does not bear the load — the frame does, but the brace prevents lateral movement. The piece of angle iron also serves as a neat mounting location for the club plaque.*

*This early rod is "sumthin' else." Check out the wild paint job, the sturdy trailer hitch, nerf bars, lantern style tail and license plate light and the Corvette suspension.*

# Wood for Rods

Most rod builders feel far more comfortable in working with wood than with metal. For the most part it is more forgiving, lower in cost, is easier to work and — for the average guy building a rod — all of the tools it takes to work with wood are readily available. Wood has a place in just about every rod we've ever seen and with some patience and common sense, wood can be used to solve a lot of rod building problems.

## Working with Plywood

Because plywood is so versatile and readily available for solving all sorts of problems, we'll concentrate our discussion on this material. There are some differences in plywood you should be aware of before dropping by the local lumber yard and parting with a twenty dollar bill or two.

One grade is called marine plywood and that is the only grade you should consider using in any kind of rod construction. This stuff is put together with waterproof glue which means the plies of wood will resist separation when the wood gets wet — and this is just bound to happen sometime during the life of the rod. You'll pay a little more for marine plywood — but for usage in a rod you'll neer regret the extra money spent.

The floorboard in front, rear, and truck should be constructed of at least ½-inch thick wood but if you have a very thin, flimsy body to start with — such as a fiberglass "T" body with no structure at all to start with, don't be reluctant to go all the way up to ¾-inch in thickness. The same holds true for firewalls even if you plan to cover them with wood or steel. If you are building a seat former or door tack panel, then you can go down to ¼-inch thick for a job which requires no bending, and down to door-skin for jobs that do (see the chapter on Building A Seat Former).

## Cutting Plywood to Shape

Shaping any plywood structure to fit in a car takes more time than it does talent. You'll need a minimum of tools. A saber saw is the hot tip for fast, accurate cutting of different shapes, but if you are dealing just with straight cuts — as is the case with some floorboards, you can build up the biceps and use a normal hand saw. A circular saw comes in handy when you start making long cuts through the thicker stuff. A disk sander or body grinder can be used to knock off edges and shape clearance notches — but then so can a wood rasp and some sandpaper. A ⅜ or ¼-inch electric drill also comes in handy.

In most cases, it is not advisable to use wood screws to attach objects to the plywood — especially if the wood is on something like a firewall. You run the risk of the screws backing out or loosening. Bolt completely through the wood or use any number of threaded fastners sold in hardware stores. In the case of something heavy such as an oil filter or master cylinder assembly, use a metal plate to back up the heavy object. Bolt through the object, the plywood and the metal base. This "sandwiches" the wood and makes for a much stronger, more trouble-free assembly.

Spend considerable time and thought to making the pattern which will be transferred to the wood. You'll be time ahead by getting the pattern perfect before doing anything with the wood. As an example, let's take the case of rebuilding a firewall with plywood. Let's assume the car has really been butchered. If the outer edge of the removed firewall is in good condition, this can be used as a pattern for the new unit. If the firewall is in very bad condition or is completely missing, you'll have to make a pattern. Before getting fired up on this project, make sure the body is aligned before transferring the firewall shape to the wood. Do this by measuring from where the original wall bolted to the frame, going diagonally upward and across to a reference point on the high side of the cowl. Measure in the same manner from the opposite rail. If the two measurements are not within ⅛-inch of each other, the body is sagging or is twisted and must be straightened until the measurements are equal. This is quite critical in the case of any roadster. Coupes and sedans are simply stronger structures and don't twist as much. But, if you are working with a car from the twenties or thirties — be ready for anything in the way of misalignment.

Once the firewall opening is known to be true, hold the plywood flush atop both frame rails and mark around the cowl lip. If the wood is to fit back inside this lip, allow for that thickness when marking the wood for the cut. If the engine and transmission are in the way, you'll have to make a pattern to fit around the bellhousing before the cowl shape can be transcribed. One of the easiest materials to work with is posterboard or non-corrugated cardboard. Hold a large piece of this pattern material against the uppermost point of contact with the bellhousing and equal height above each frame rail. With a pencil, trace the outline of the bellhousing onto the cardboard. Work from both sides — going toward to the center. Keep cutting the shape away and remarking until you have a pattern board that completely clears the bellhousing equi-distant all the way around. A pair of scissors, an X-acto knife, a pencil and some patience is all you need to do a good job here. When the pattern fits the car perfectly, transfer it to the plywood.

Cut the plywood, set the panel in place and then mark the sides and cowl outline. Don't make the mistake of putting the bottom edge of the firewall too close to the engine or bellhousing. Keep in mind there is always some movement between firewall and engine. Fiberglass matting or cloth and laminating resin can be used to attach the firewall to the cowl on the driver's side. Make sure you have the firewall placed just where you want it before starting this phase of the operation. Once the firewall is glassed in place, it's a good idea to coat it with a mixture of resin and hardner on both sides and the bottom edge to completely seal out all moisture.

Floors, seat bottoms, and upholstery tack strips are just some of the items found in street rods that can be constructed of plywood. Glassing them together or in place within the rod is the accepted way of putting everything together and sealing out all of the dirt and grime.

## Other Woods

In the case of decorative and functional items — such as running boards and instrument panels — you'll want close grain boards such as oak, ash, maple or mahogany for a beautiful finish all the way around. If you don't have access to a router or to some of the furniture quality woods we've mentioned, drop by your local cabinet shop for some help and advice on where to get what in the way of wood in your community.

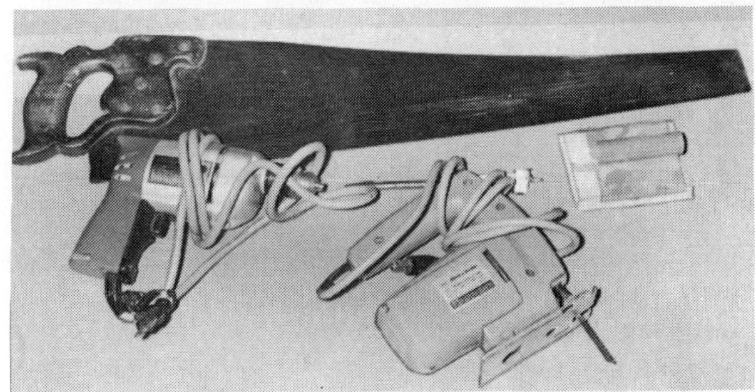

*Woodworking for rods requires very little in the area of tools. To completely flush out the assortment shown here would be a router, disk sander and rasp.*

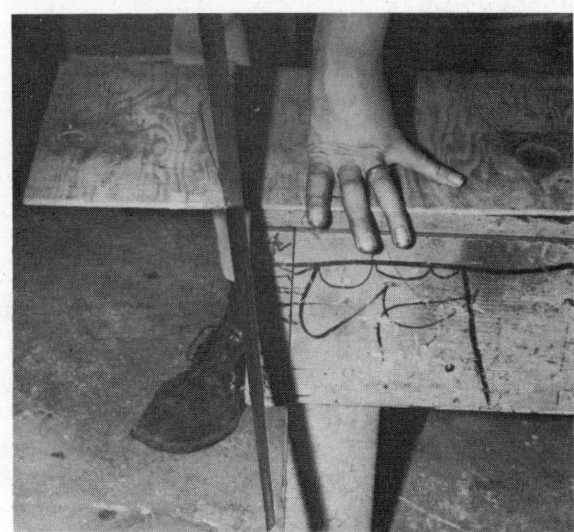

*More than one Model A floorboard has started life right here — after forty years, this is still the low cost, simple way of taking care of the problem.*

*Competition style reproduction fiberglass bodies have no built-in structure and no floor. Plywood is the only sensible answer.*

*A non-corrugated cardboard is the hot tip for making a pattern for any kind of wood use in a rod, but cut open boxes, and even newspaper can be pressed into service to make a pattern.*

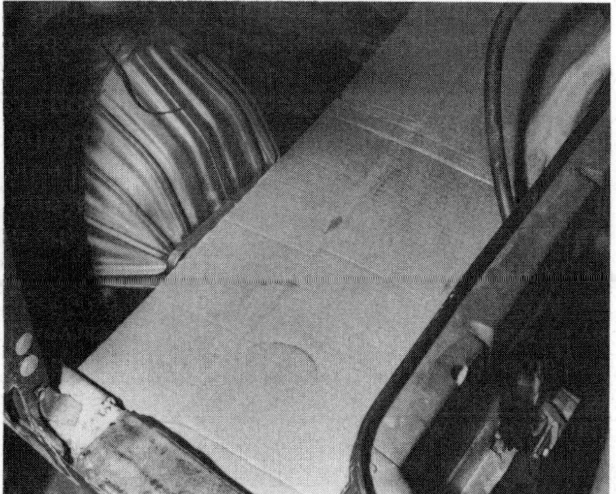

*Regardless of the material used for the pattern, at this point it's a matter of cut and try until you get it perfect. Time spent here is gained further down the line.*

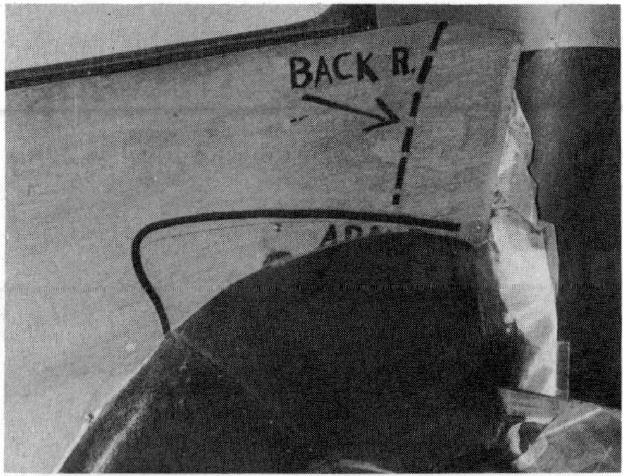



*Here's a good example of door skin being fitted into the back quarter panel of an early sedan. Inner fender panel has been glassed for additional strength to a rusting body. Waxed paper serves to keep the mess to a minimum.*

*If you need to go around turns, or build up laminates, door skin is the ideal material. Soft, thin material can easily be cut with aircraft shears.*

Here's an almost classic example of wood being used in modern rod construction. The ³/₄-inch plywood floor is glassed into the lower edge of the thin fiberglass Model T repro body. The basis for the seat former has already been put in place.

Floorboards and toe boards in early rods can be glassed together at the appropriate time in the construction of a rod and then covered with resin and hardner (even cloth) to make for a virtually indestructable assembly. Just make sure everything is the way you want it before glassing it all together.

(Right) Don't rush wood. When building something of wood that gets glued, or laminated, plan on plenty of time being spent on making patterns, and getting it all clamped together.

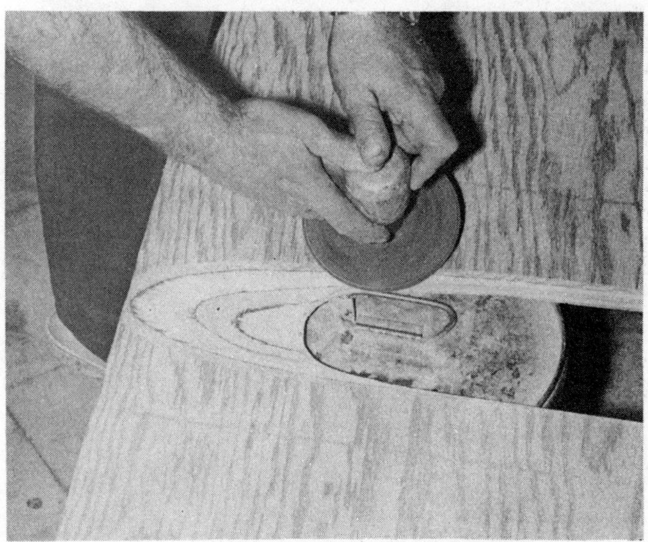

A disk grinder or sander can really speed up the shaping process when wood needs to be contoured . . .

. . . to fit over a rounded shape like the raised driveshaft tunnel in this chopped and channeled '48 Ford coupe.

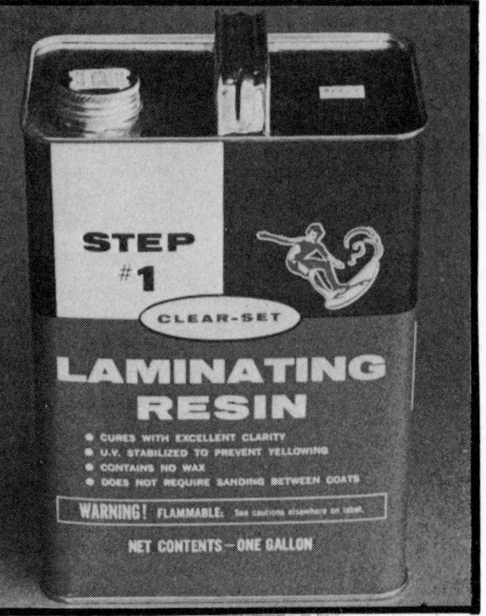



*The binding agent for fiberglass is available in mat or cloth forms; both are bonded to the working surface with laminating resin and hardner. See your local boat repair shop or large paint supply store. Properly used, this combination can be the amateur rodder's salvation.*

*Rough-sawn exterior plywood is the "in" thing in exterior construction and decor in many part of the country. Obviously, it is ideal for use on the exterior of many a street vehicle.*

*The hardware, the usage, the shapes, forms, textures, and colors of wood all go hand in hand to make this a most useful tool for the rodder.*

*At first or second glance this instrument panel is not wood — but just keep looking, it is — an excellent example of what can be done.*

# Building
# A Seat Former

Building a seat former — or a base — is the first step in building a fixed seat of the type found in a roadster. The method is essentially the same regardless of whether the body of the car is steel of fiberglass. Construction of the former is relatively easy and low cost, and most rod builders can handle the construction with no problems.

The following photographs outline the procedure for not only installing the tack strip and seat former, but also how to go about laying out the design for button tufting upholstery.

*The tack strip goes around the upper perimeter of the seat area of the body. The construction and attachment of the tack strip should take place before any work is done. A number of pieces of hardwood (oak, ash or birch) should be shaped to mate with the upper lip of the body. It is essential that the tack strip be at least ⅝-inch thick and extend out from the body by a like amount. This allows the use of an upholstery tack which has a length of ½-inch. The tack strip can be attached to the body with countersunk, flathead screws, lock washers and nuts. In the example shown, the rod has already been painted. From a practical standpoint, it would be wise to install the tack strip and construct the seat former before painting.*

*The seat former is made with layers of door skin. This is a low cost, flexible, ¼-inch thick laminate — not to be confused with plywood. Door skin is much more flexible than plywood. Determine which way the door skin is most flexible and work it to your advantage. The material can be soaked with water to make it even more flexible. Jam it in the body and hold it to the lines of the body firmsly with braces and clamps. Build up at lease three layers of the door skin — bonding the layers together with a water soluble wood glue. Do not use contact cement! Misery is guaranteed if you do! Noted that the platform for the seat bottom has been shaped (by trial and error) and put in place on the seat riser in order to help hold the lower portion of the former in place.*

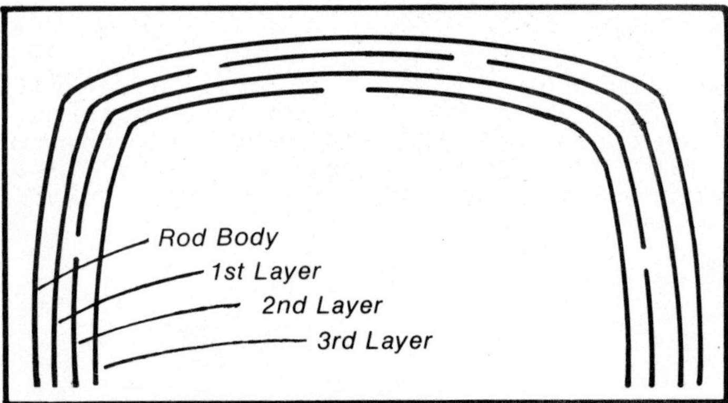


When laying up the three layers of door skin, make sure that the seams do not overlap. The door skin is so easy to work with that a template need not be made. Rough dimensions can be transferred directly to the skin and the preliminary cutting initiated.

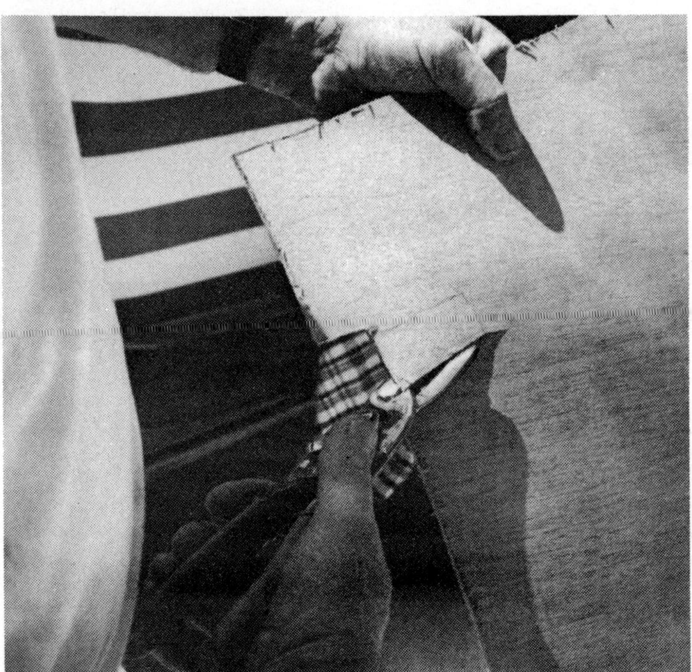

Ordinary tin snips or compound shears can be used to cut the single layers of door skin to fit. Saber saws with a fine blade can also be used, but the tin snips seem to be faster and more accurate.

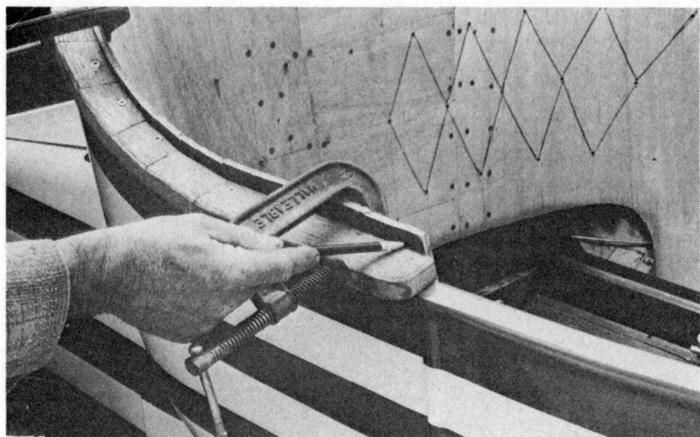

The final step in the actual construction of the former is to mark and then trim the upper edge of the former flush with the upper edge of the tack strip. Note how this tack strip is made up of a number of small sections of wood. At this point it is important to note that the seat former should remain clamped in position for a minimum of twenty-four hours after the last layer is glued in place and that it is not attached to the tack strip — only clamped there. The seat former must be constructed so it can simply be lifted from the car once the clamps have been removed.

One of the most critical steps in upholstering with button tufting is to achieve the effect you want with the initial layout. The layout can be done simply by measuring and drawing out what you want onto the seat former. A template (the cork diamond shown here) is preferred by others to establish the pattern. It is important that the layout be started in the center of the seat former — both vertically and horizontally. The density of the buttons and the shape of the pattern is a matter of personal taste.

(Left) While the final layer of door skin is drying, the layers of door skin should be attached to one another by nails, moly bolts or in some cases — small butterfly nuts. It is imperative that the layers of door skin do not separate from each other. Some of the dots showing on the door skin in this photo are of nails used to help bond the layers together and others are layout holes indicating where button will be placed.

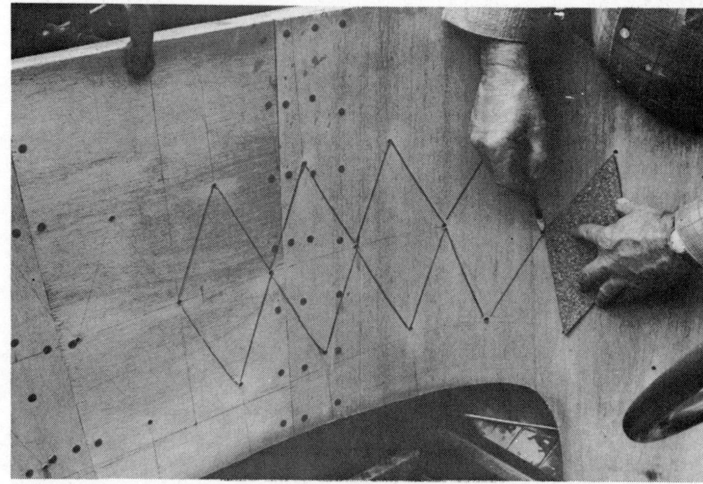

*(Left) complete the layout before starting any of the upholstery. It is wise to mark off the initial layout with light pencil so the pattern can be easily changed. Keep in mind how the pattern winds up — at the top of the seat and at the ends where the door meets the former is very important once the project is finished.*

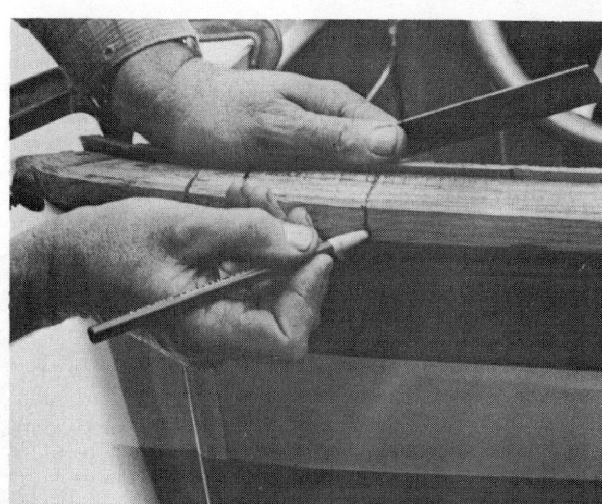

*It is most important to finish the layout on the former down to the last tuck and fold if you are new to upholstery. A professionalf can visualize a lot of this and experience will keep him out of trouble, but an amateur would be wise to follow through on this layout procedure.*

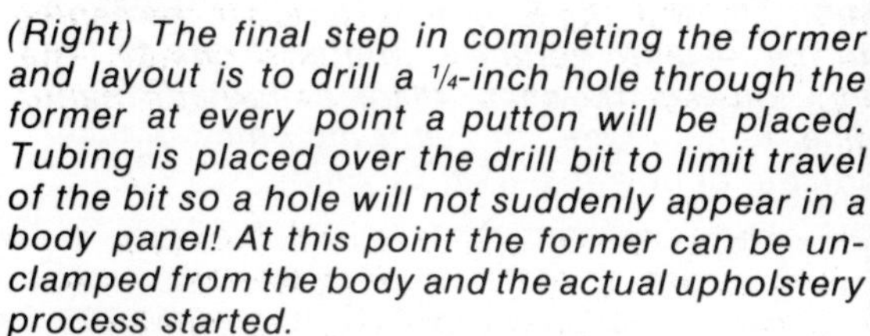

*(Right) The final step in completing the former and layout is to drill a ¼-inch hole through the former at every point a putton will be placed. Tubing is placed over the drill bit to limit travel of the bit so a hole will not suddenly appear in a body panel! At this point the former can be unclamped from the body and the actual upholstery process started.*

# Rod Building Tips II

This small block Chevy is one clean engine. Note how much the polished aluminum water pump and manifold do to accent the ribbed valve cover. Routing of plug wires would also be difficult to fault.

The alternator needed to be tucked in quite close to the crank on this Olds engine which was shoehorned into a mini-truck. That home-made alternator mounting bracket is a good example of how it should be done.

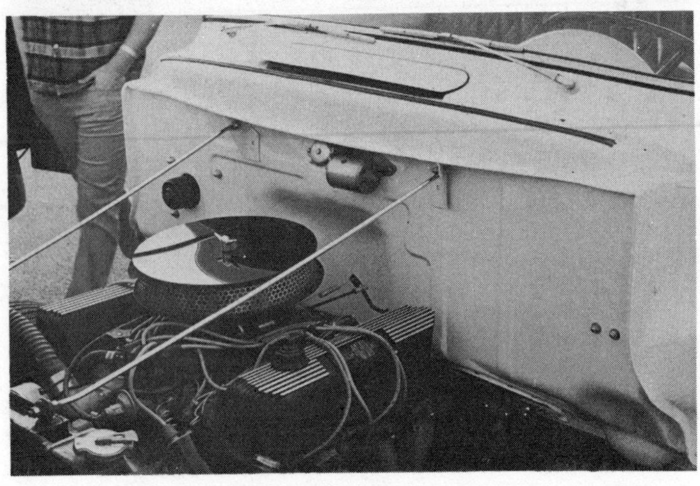

(Left) The firewall on this '65 Ford pickup is just about as clean as it could be and still retain the stock wiper and heater motors.

Here is another very fine example of how a small block Chevy can be detailed for rod use. Note how the plug wires have been routed under the headers and the water temperature sending tube has been laced to the wiring leading from the alternator.

This is an excellent example of "do it yourself engineering." This rod owner scored the use of available air cleaner/scoop arrangements and fabricated this gem. Neat, huh?

Here's a rod that belongs to someone bound to have fun — rain or shine. With "Joy Ride" on the cowl, a trumpet on the windshield post and a small block Ford up front, what else would you expect from a T-bucket?

This sort of project is just perfect for the enthusiasts who live north of East Overshoe, Montana. Keep in mind that when the garage is snowed in, you can work on the rod project one step at a time at the kitchen table.

This ain't your ordinary J.C. Whitney, K-Mart, Western Auto battery box hiding under the floor of this Ford Phaeton. You'll note that the frame is not exactly stock either.

Nothing fancy or exotic here, except that you don't see a fully chromed, firewall-mounted master cylinder very often — and it sure does look neat.

*Fortunately for all rod builders, there is a great abundance of hinge mounted mirrors. They can be had in most any size, shape or distance from the hinge.*

*Here's a prime example of a clamp-on hinge working well on a forty or fifty vintage car. Nice.*

*There was a time when this type of mirror and mounting was as popular as fender skirts and fox tails.*

*Mirrors that clamp in place with set screws are as readily available as the more popular early car hinge-mounted mirrors. The doors on this car hinge from the rear, so the sun visor is an ideal location for mounting.*

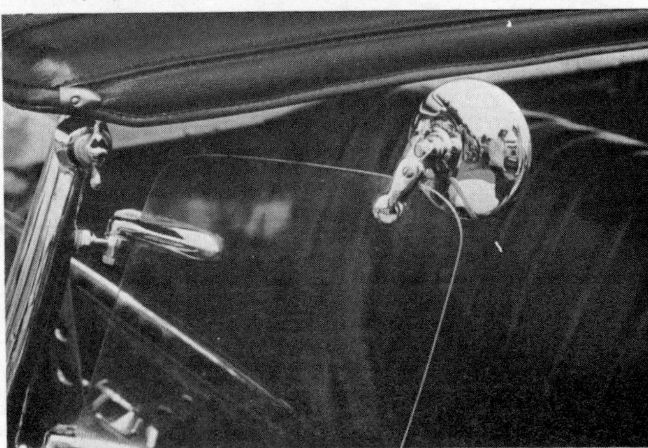

*Not many cars can utilize a glass mounted mirror. Glass shops can drill the necessary hole without strain or pain. Current technology allows mirrors to be epoxied into place.*

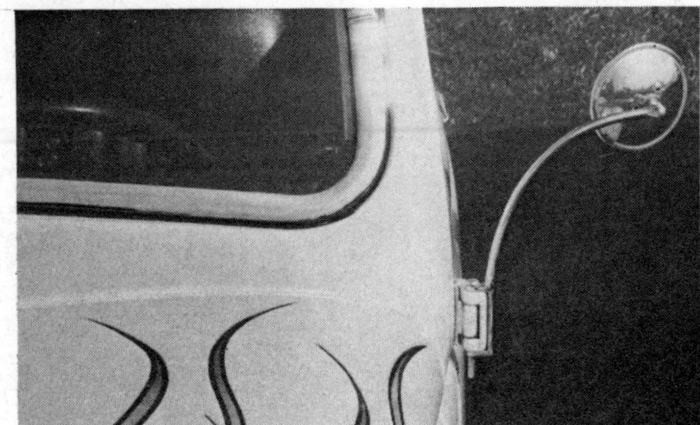

*This closed car features a hinge-mounted mirror which was once very popular. Swap meets and your own fabricating ingenuity come in plenty handy here.*

A padded panel such as this will always give a rich, custom look. Check the very subtle striping accents just below the windshield.

Mirrors mounted on the hinge are quite effective when the placement is above normal eye level.

(Right) This is a Stewart Warner panel intended for boat use. Obviously it is right at home in this well-turned-out roadster.

(Below, right) This is the stark, functional look of hte late forties and early fifties. If that is the look you are after, then study this photo long and hard.

(Below) The extensive use of engine turning goes back more than fifty years. It still gives an aura of quality craftsmanship.

Cable actuated throttles are simple, clean, easy to work with and almost never give any trouble. This home-made setup is about as simple and straightforward as they come.

Most all late model domestic passenger cars (and a lot of foreign cars) have shielded throttle cable setups that are easy to adapt to a rod.

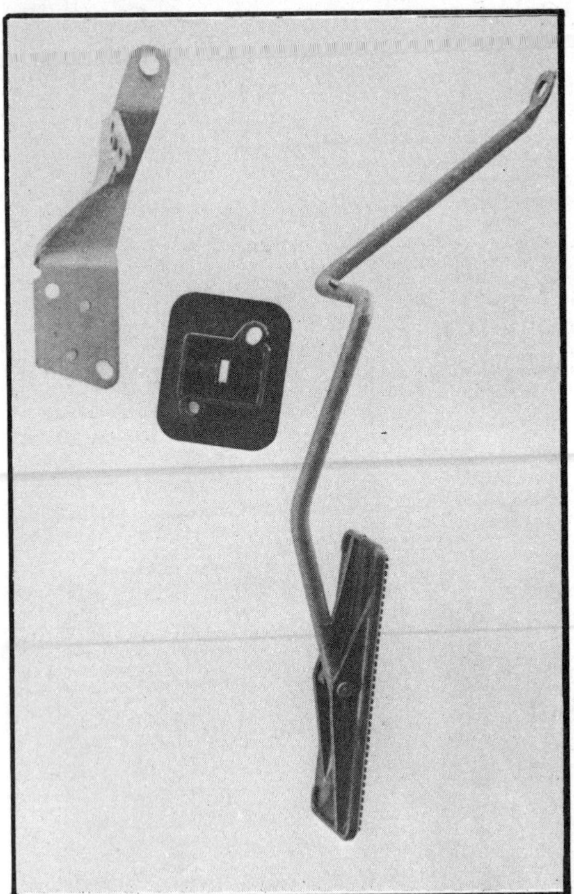

On a throttle cable setup this is all that is needed on the inside of the car. Arm of throttle pedal sandwiches between the two brackets which in turn bolt to the inside of the firewall.

Using a machinist's level when installing an engine in a rod insures that the block is level from side to side.

If your rod happens to be a pickup truck, consider this "double deck" arrangement which adds greatly to the amounnt of stuff you can haul with ease on rod run trips. Carpeted upper deck makes a nice place for rug rodents to take a snooze.

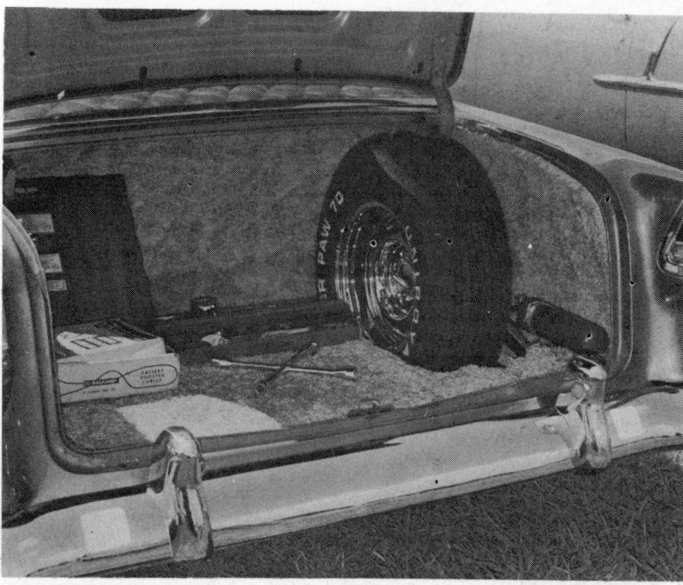

Plush carpeting and a little bit of chrome plating can go a long way in detailing the trunk of any rod.

Check the gas cap, CB antenna, power seat and chromed bed brace on this Ford pickup for detailing.

Light colors, plenty of instrumentation and toggle switches were all the rage in the fifties. Keep it in mind if you are building an era custom.

The bed of this pickup is detailed to the "Nth" degree with cedar, chrome carriage bolts and close-cropped carpeting along the sides.

The imagination of a good rodder knows no bounds. In this case the grille and headlight assembly from a late model Ford sedan finds a home in this late model Ford truck. Trick, huh?

This is more than just a little space age electronic wizardry attached to this digital readout, push button dash. Don't overlook that sports car steering wheel — a nice touch.

The dash on this coupe may look plenty fancy, but on closer inspection you'll find there is a lot of function here and those large wood panels just might serve to hide a multitude of previous sins committed to the dash.

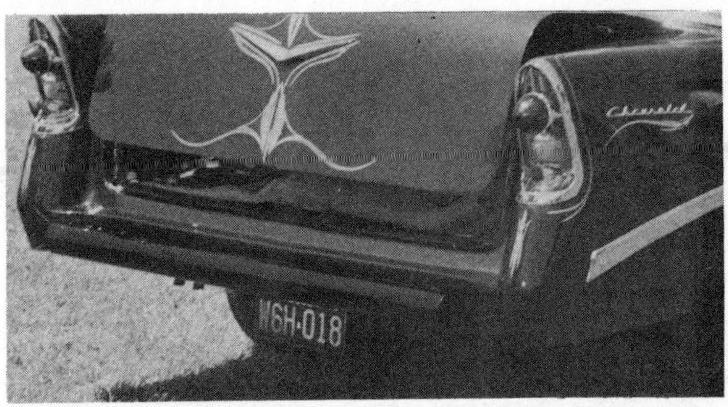

A variety of effects can be achieved with a custom bumper — in this case made of round tubing or pipe. On this '56, the lower trailing edge of the fender had to be reworked to finish off the project.

We don't care if the crank is functional or not, in the case of the braided leather strap it is the thought that counts.

If you would rather not mount your rod run dash plaques to the dash, consider an arrangement such as this. Obviously a mounting board such as this can be removed and placed in your "new car."

# Button Tufting

Traditionally, one item brings the backyard rod builder to his knees asking for help from the professional. That item is, of course, upholstery. Rod builders will spend hours beating out dents in a fender, block sanding, masking and painting — have the job turn out bad — grind it all off and start over again, but they won't begin to tackle the job of doing an interior. A certain aura of mystery surrounds a custom interior and most of us duck the issue with a single statement: "I don't have a sewing machine."

There is one style of upholstery which can successfully be done without a sewing machine and it can be handled by most anyone with a few hand tools, a little forethought and some patience. This is button tufting. Tufting can be done with a fabric or with naugahyde. It can be done in a diamond or square pattern, and it can be done at home in a "T" or in a full bodied custom. Tufting is equally at home on a "built from scratch seat" or a stock rebuilt seat.

The layout is everything in tufting. The basic mechanics of tufting are very simple and routine, but the tufted seat is not worth installing if the layout has not been carefully thought out. All alyouts on seats or seat backs must be worked from the center outward. Trace the outline of the seat on the garage floor, on butcher paper or on an old sheet. Regardless of the method you choose, have it set in your mind exactly where all of the buttons should go long before you pick up one piece of material.

By doing this you make certain all buttons will go just where you want them on the finished product.

If you are starting with an essentially stock seat, the preparation is relatively simple. The springs must be covered with burlap which is stapled or laced to the spring frame. To give body and firmness to the seat, the burlap should be topped with firm, bonded foam three inches thick. On top of this goes a two-inch thick slab of urethane foam — provides a softness found in the most expensive luxury cars. The two foams are bonded together with contact cement which is available at the local hardware store. Once the foams have been bonded together and placed atop the burlap-covered springs, the padding can easily be trimmed with a household electric carving knife. If you wife objects, a razor blade can also be used — it's just slower.

At this point you should have the layout of the tufting well in hand. If not, keep working on the layout. Start from the center of the seat and work outward following your own dimensional pattern to the edge of the seat. A 4½- to 5½-inch square pattern is very common in street rod upholstery.

Once the layout is completed, it can then be transferred to the top layer of foam on the seat. Think about it a long time before you mark it — you'll have to live with it a long time, or scrap it and start all over. Never lose sight of the simple fact that in tufting, the layout is everything.

In transferring the layout to the foam, simply make an "X" at each point you plan to have a button. Now you are ready for material selection. Either naugahyde or fabric can be tufted — but for the novice, fabric is easier. When selecting fabric, avoid material with a corduroy-type grain. This can be used — but you shouldn't do your first tufting job with this type of material. When selecting a naugahyde to be used on a tufted seat, make sure you choose one thin enough to hold a fold — since a fold is an essential part of the tufting process. An additional tip is that you shouldn't attempt to tuft naugahyde on a cold day when the material can be quite stiff and difficult to work.

In order to begin tufting, you'll need covered buttons—presumably covered in the same material you have selected for the seat. Most any upholstery shop sells buttons and can cover them for less than a quarter a piece. Just decide how many you need and provide the shop with the fabric. At the shop you'll be asked what size button you want — since there are several sizes. One of the most popular is #30. You can stick with that or make your choice from the selection offered by the shop. With the layout completed and the buttons and fabric in hand, you are ready to begin tufting.

At every point you have marked an X on the foam to receive a button, a hole must be punched. The hole needs to be about 3/16-inch smaller than the outside diameter of the button. A punch can easily be made by selecting a short length of tubing having the correct inside diameter. Sharpen and taper the end of the tubing on a bench grinder so the ID becomes the cutting edge. Lay the foam on a board and drive the tubing downward to punch the hole. It is quite important that the tubing go straight through the foam — so be sure you are holding the tubing at right angles to the surface of the foam.

After all the holes are punched, the foam can be placed back on the burlap-covered seat. At this point the layout needs to be marked off on the backside of the fabric to be used. This layout will be different than that on the foam. The layout will have the same pattern, but different dimensions. In the case of a five-inch square pattern on the foam, the layout on the material is set at six-inch squares. This "spreading" of the pattern is necessary to provide enough material to draw downward with the buttons. The rule of thumb on "spreading" the pattern is to add an inch to the distance between the holes.

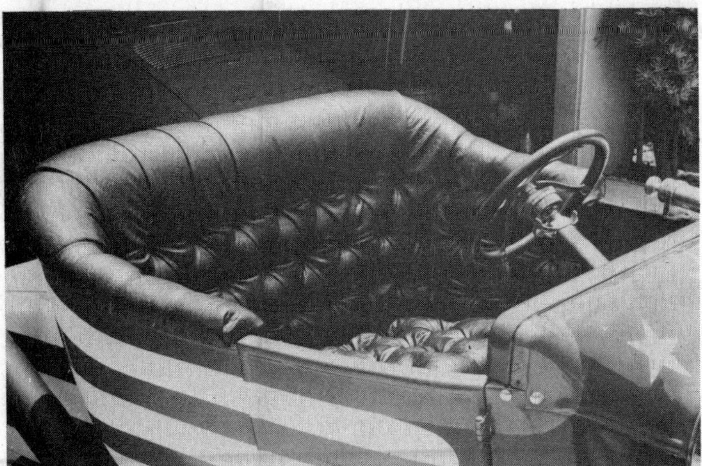

*This very handsome button tufted interior for a T-roadster was done without access to a sewing machine or any other special tools. The tufting can be in either fabric or Naugahyde on either spring type or "solid" foam rubber type seat construction.*

*Spray adhesive can be purchased in hardware or paint supply stores. If a foam seat is being built, this adhesive is adequate for bonding the layers of foam together.*

*Buttons are readily available at upholstery shops in a variety of sizes — one of the most popular is #30. Before buying the buttons you should determine exactly how many you need — determined by the layout — and the material you will want the buttons covered with. Upholstery shops have a neat little gizmo that will cover buttons in the material of your choice for about twenty cents a button — plus the material.*

Like the layout of the pattern, tufting should begin in the center of the seat — then outward to both ends. Make a large needle from a piece of welding rod — sharpen one end, then flatten the other and drill a hole in the flat portion large enough to accept the cord attached to the button. Pull the button down to the top of the material using the needle, then run the needle through the hole in the foam and then through the burlap cover over the seat springs in order to bring the button cord out in the spring area. Once the cord shows on the underside of the seat, the needle can be removed. Shape a "bowtie" or "dog bone" from a small wad of cotton. This will be placed on the underside of the seat in the spring area, and the button cord is tied around it in order to keep the button pulled down tightly against the fabric and the foam. Now a button knot can be tied in the cord and pulled down tightly against the cotton wad. This in turn will sink the button into the fabric. Don't get carried away with this — a button knot can always be tightened, but can never be loosened. You will want all of the buttons to be sunk to the same depth in the seat so plan on going back over all of the cords after the seat is finished — several times — keeping in mind you can pull a button down another quarter of an inch, but you can't bring it back up unless you cut the cord and then resink the button from scratch.

The button knot is essential to button tufting and that's why we've included a separate section on just how to tie the button knot. It is that important.

As you begin pulling the second button into place — and the third and fourth and so on — folds will begin to naturally appear between the buttons. Before pulling the button down tightly, even out these folds or tucks with your fingers or with a thin, non-cutting instrument like the backside of a table knife. You will want all of one kind of folds to lay in one direction. This is easy to do so long as you think about doing it while the buttons are being pulled down.

Once all of the buttons have been pulled down, the material can be brought on around the seat and attached to the underside of the seat with a staple gun or with small tacks. A third method is often used on newer seats — that of using hog rings to attach the material to a heavy wire running on the underside of the seat.

Although we've run through the process of tufting a stock seat, the process is essentially the same for a scratch-built seat. Use a plywood base, cover with foam, lay out the pattern on both plywood and foam, cut the holes in the foam and drill ¼-inch holes in the plywood at each point there is to be a button.

Button tufting has everything going for it — quick, easy to do, low in cost and you don't have to have a sewing machine to do any of it.

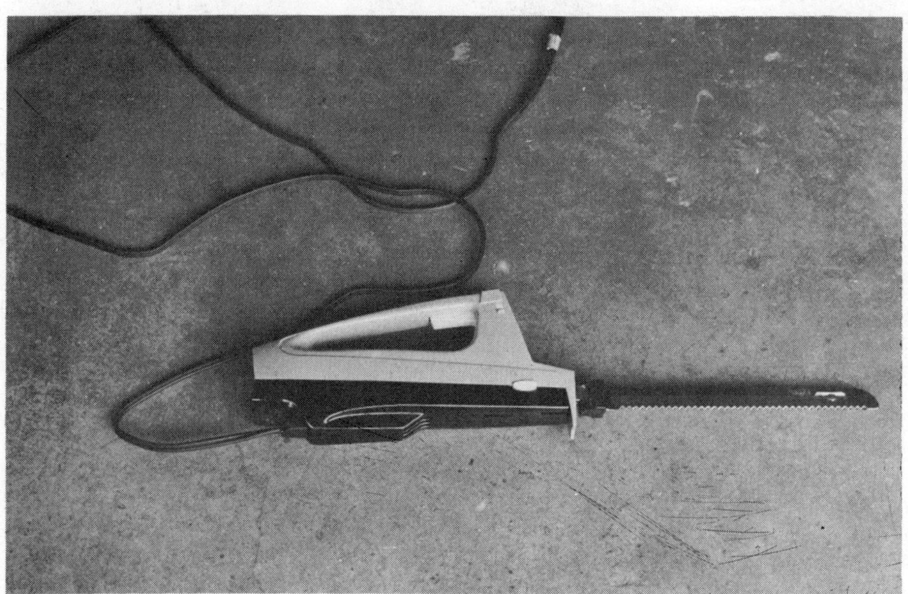

*Foam for upholstery is most easily cut with an electric carving knife. Alternatives are X-acto knives and razor blades.*

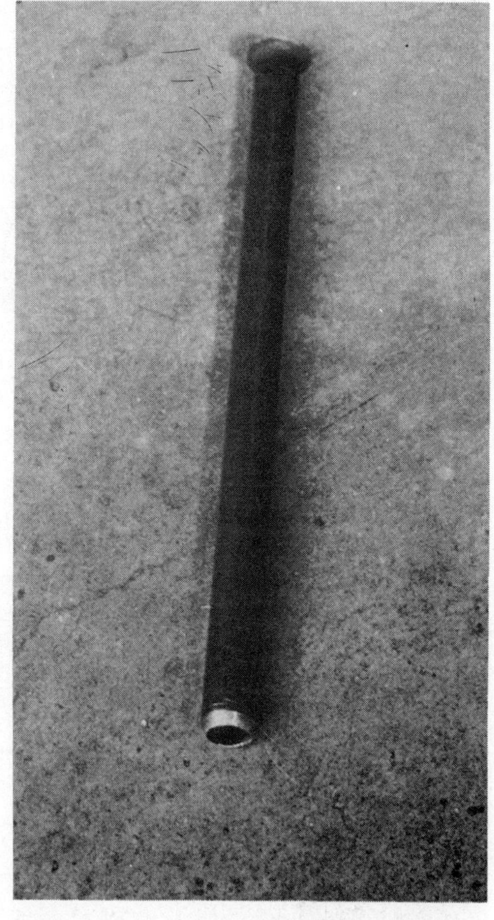

*If you are to be button tufting over foam rubber you'll need to make a suitable hole punching tool from a short length of 3/16-inch ID steel tubing. Simply sharpen one end of the tubing with a file or bench grinder and the tool is ready to use.*

The tufting pattern to be followed on the seat former needs to be transferred to the backside of the fabric to be used. The pattern on the cloth needs to one inch larger at every point than on the seat former. In the example shown, the cork diamond template was used to layout the pattern on the former, while the cardboard template underneath will be used on the fabric. This "spreading" of the pattern is necessary to provide enough material to draw downward with the buttons.

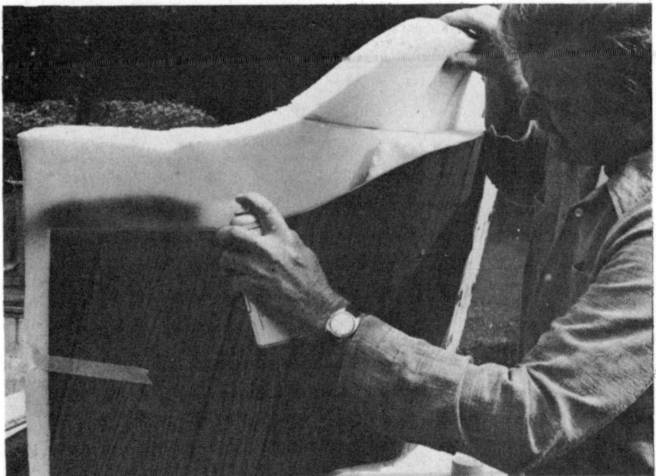

One fo the quickest and most accurate ways of fitting the foam to the seat former is to tape a section of foam to the inside of the former and spray paint along the edge to achieve a cutting outline. Paint sprayed through the holes in the former also locates the holes on the foam.

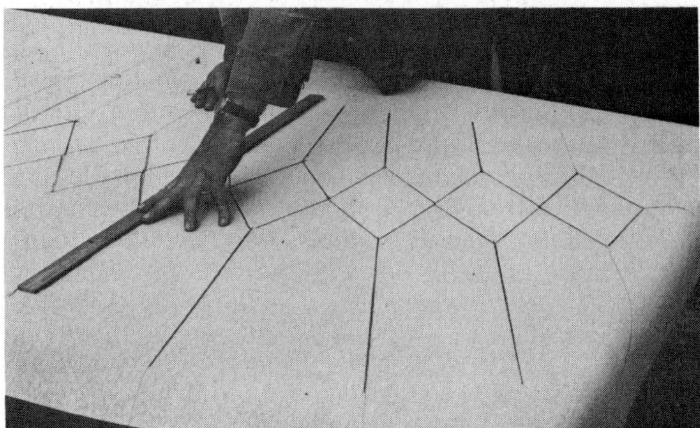

In the case of a curved seat back former, the pattern on the back of the fabric must be curved or "splayed" to match that of the former when the two are joined. This is best done by transferring the former pattern to butcher paper when the seat former is in place, and then straightening the paper for easy transfer onto the fabric. Keep in mind that at this point the button location points still must be "stretched" as previously described.

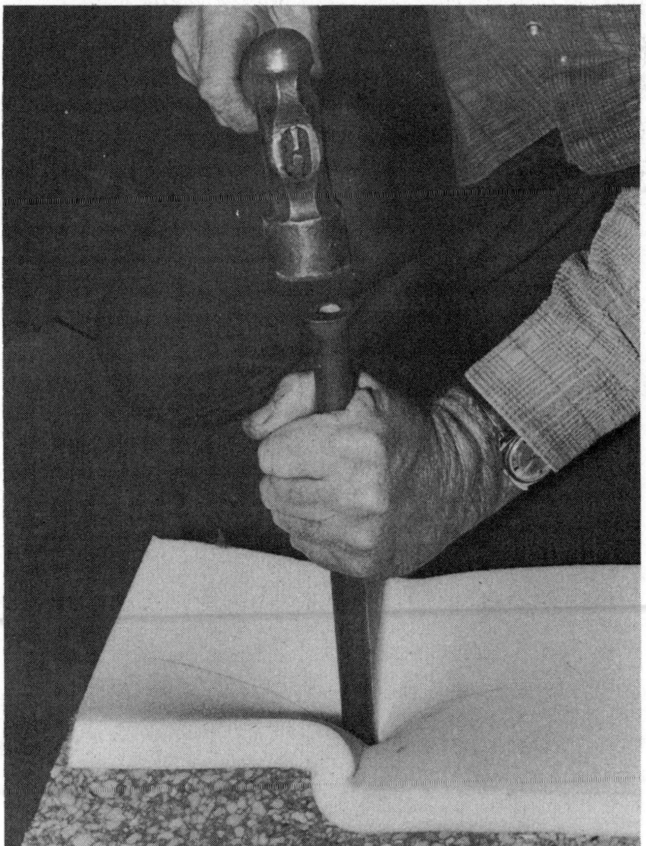

After bonding the two layers of foam together, the punching tool can be used to stamp a hole in the foam at all button locations. If the tufting is to be placed on the seat (as opposed to the seat back) then a three inch layer of bonded foam should be used in addition to a two inch layer of urethane foam.

If a seat back is being covered with urethane foam, then two two-inch layers may be glued together to provide necessary support. The electric carving knife is an excellent tool for any cutting that needs to be done on the foam.

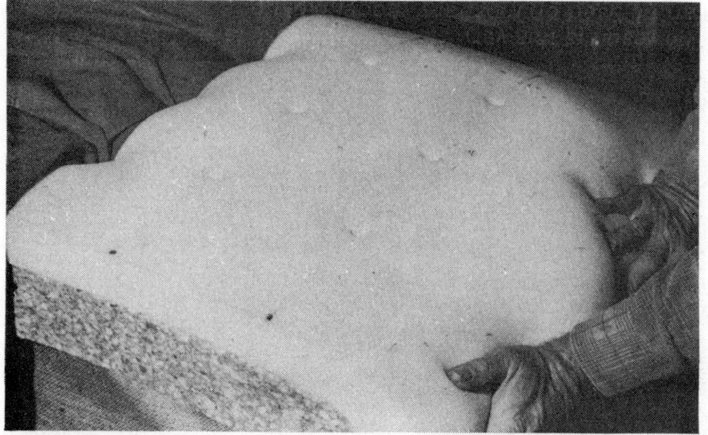

*Don't try to rush the project. When bonding the two foams together, make sure the glue has set up before punching the holes. When punching the holes, make sure they reflect the accurate pattern you are absolutely sure you want. The burlap showing beneath the bonded foam in this picture will separate foam from the springs on a stock type seat. This must be done to keep the springs from "eating" through the foam. If the foam is to be placed over plywood, then burlap is not needed.*

*(Below, left) The first step in installing a button is to thread both ends of the button cord through the large needle made from a section of welding rod about 8-inches long.*

*(Below) With the pattern marked off on the backside of the material, insert a probe or needle in the center of the X in order to correctly locate the button on the other side.*

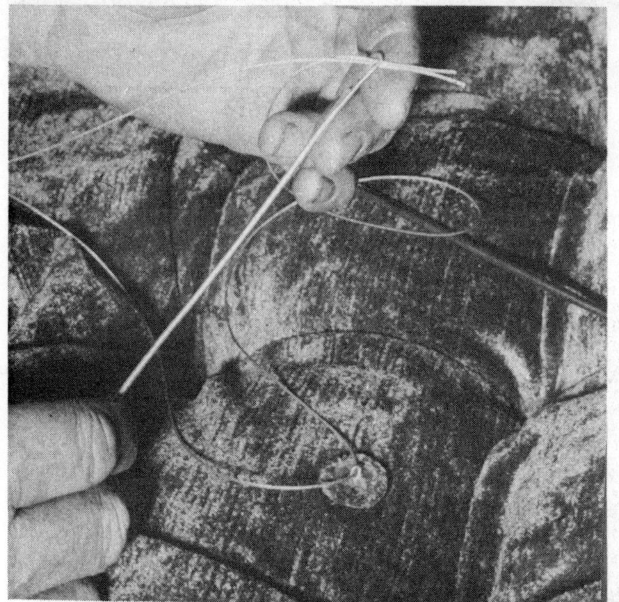

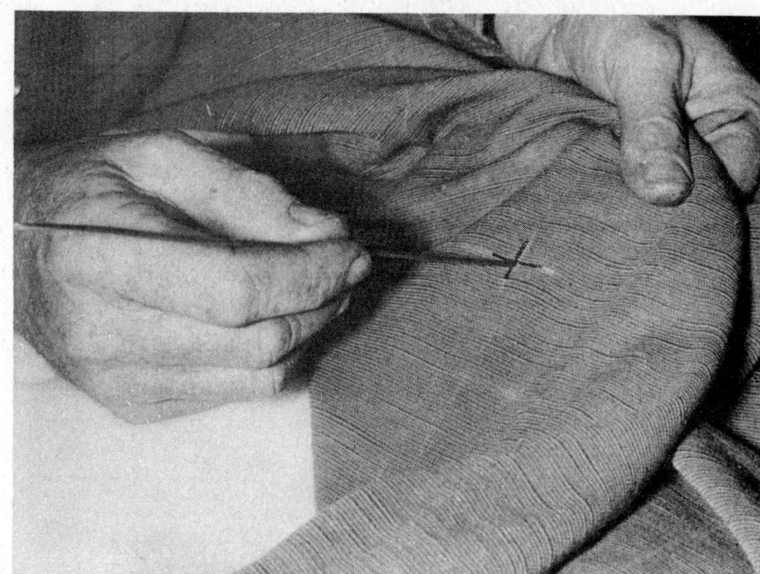

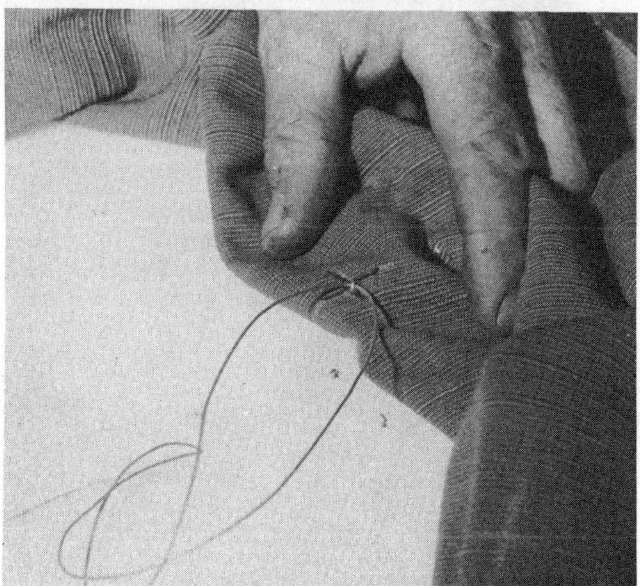

*Push the needle and the cord through the hole in the material and work the bottomside of the button into the material until the steel eyelet comes through on the bottomside of the upholstery material.*

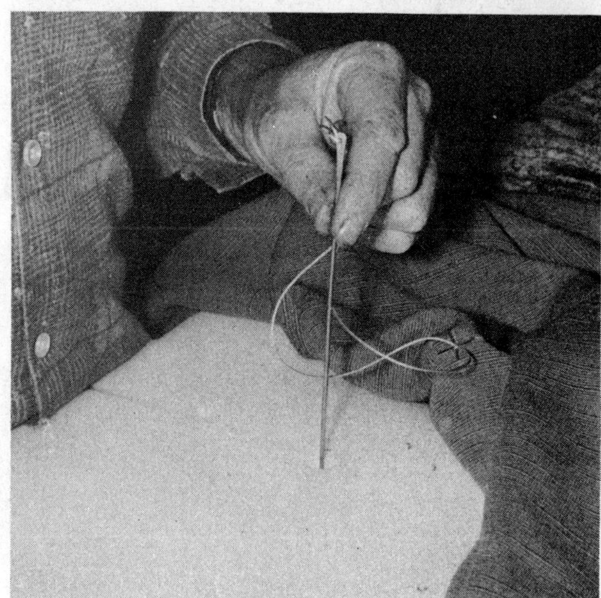

*At this point the needle can be dropped through the hole in the foam and burlap and into the spring area — or in the case of a foam and plywood seat — through the foam and plywood.*

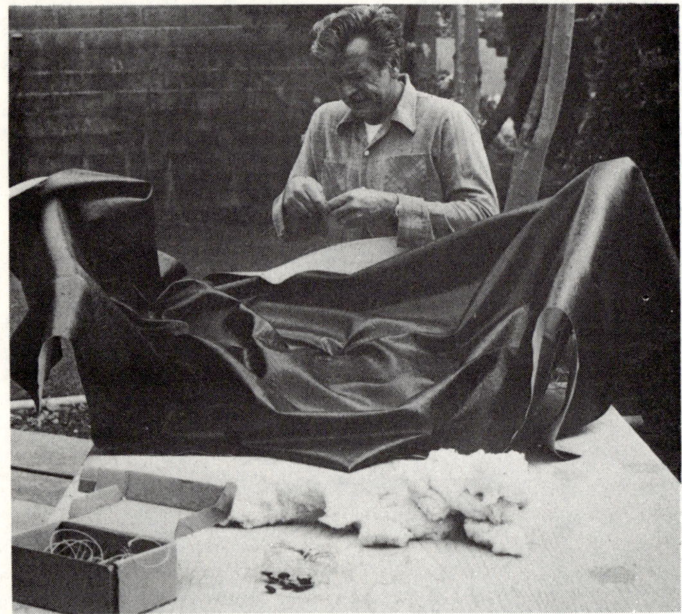

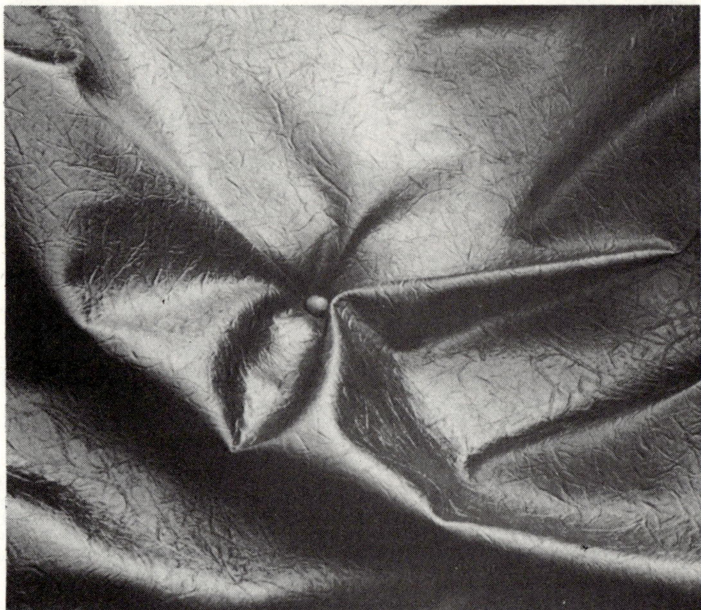

Seats and seat formers are bulky items and at times you'll find yourself working on "both sides at once" so plan to do your upholstery work in a well lighted, spacious area. Note that in the case of the seat former shown here, a quite large piece of naugahyde is being used. If it is too large you simply trim it. If it is too small you start over.

Regardless of the project, it is essential to start in the center and work outward. Do not try to pull the button down to the full depth at this point. Increasing the depth of the buttons can always be taken care of later.

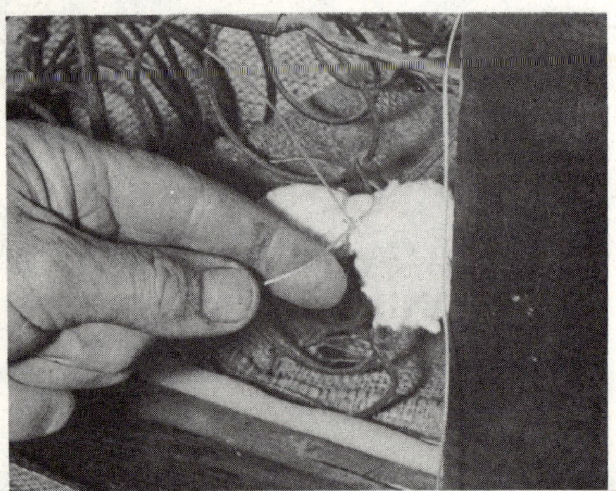

(Above) After the cotton bow tie is in place, begin tightening the knot which will shove the cotton toward the plywood or burlap. On a spring type seat a long screwdriver with a small notch ground in the end of the blade makes an excellent tool for maneuvering around the springs.

(Above, left) Once the cord has been pulled through the burlap or plywood, fashion a small bowtie from a wad of cotton. Tie a button know and place the cotton between the knot and the plywood or burlap.

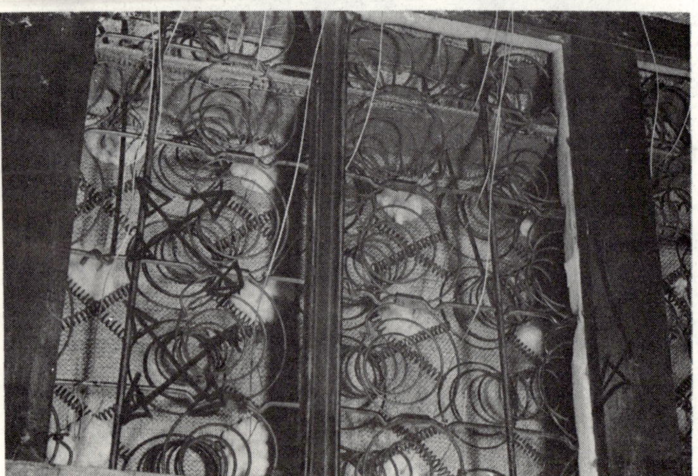

(Left) The backside of this '40 Ford seat shows a wad of cotton at the point of each button on the underside of the seat. Note the thin layer of urethane foam placed between the wooden structure of the seat and the springs to eliminate noise.

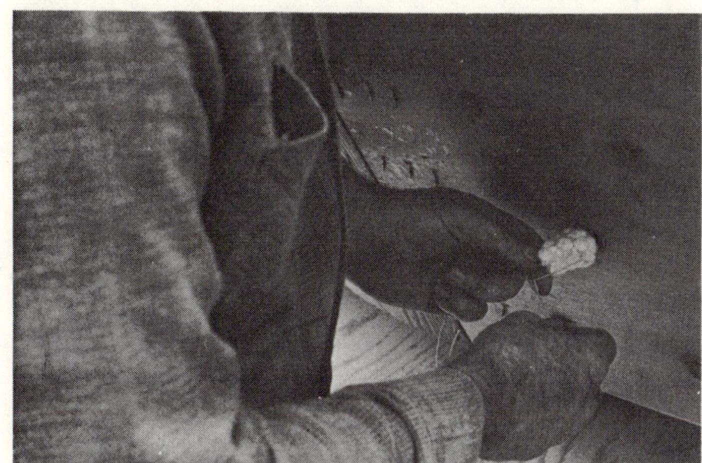

The tying operation on a plywood seat former type of seat is easier and faster than on a spring seat, but the basics are the same.

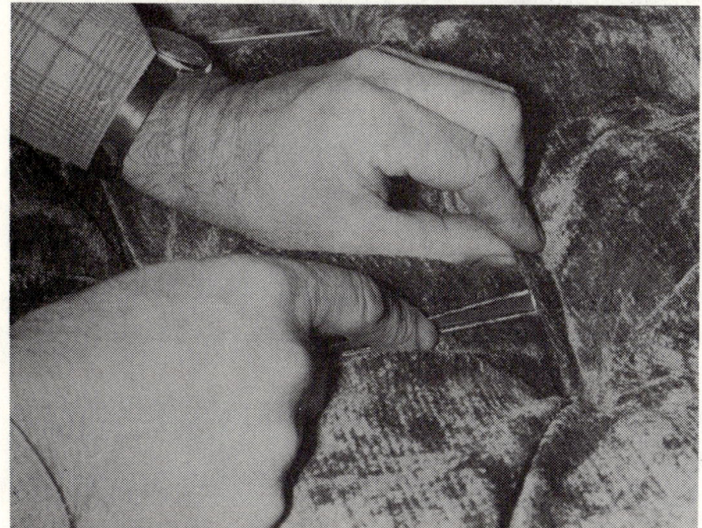

As the buttons are pulled down a fold or crease will appear between buttons. These should be evened out and made to lay in one direction before the knots are tightened.

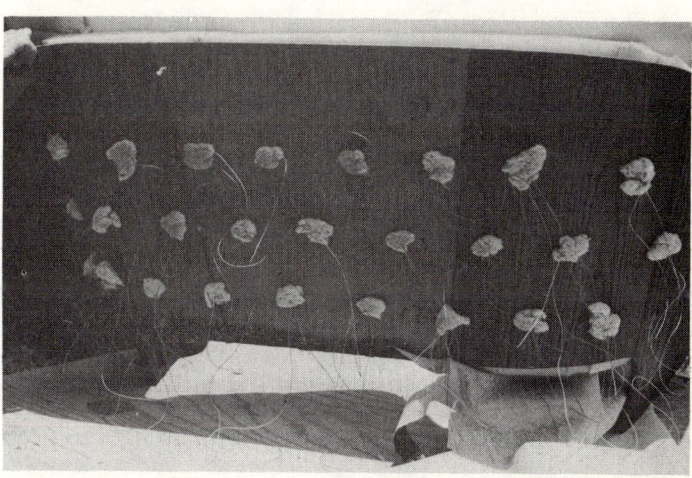

This is what the back of a seat former will look like once all of the buttons are in place. The upholstery material at the top of the seat has been folded forward for picture taking purposes. In time it will be pulled to the back and stapled to the underside of the tack strip on the body. Note how the stapling operation is going on at the bottom of the former.

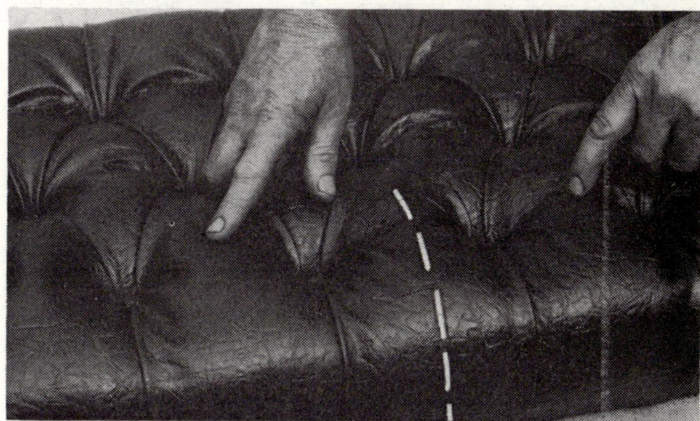

The dotted line in this photo indicates the lateral center line of the seat. Note how the creases are folded away from the centerline to provide a symmetrical fold pattern.

This seat is about two-thirds finished at this point. Note how careful attention has been paid to how the creases between the buttons are folded. Note also that at this point some amount of care has been given to pulling the buttons down an equal amount.

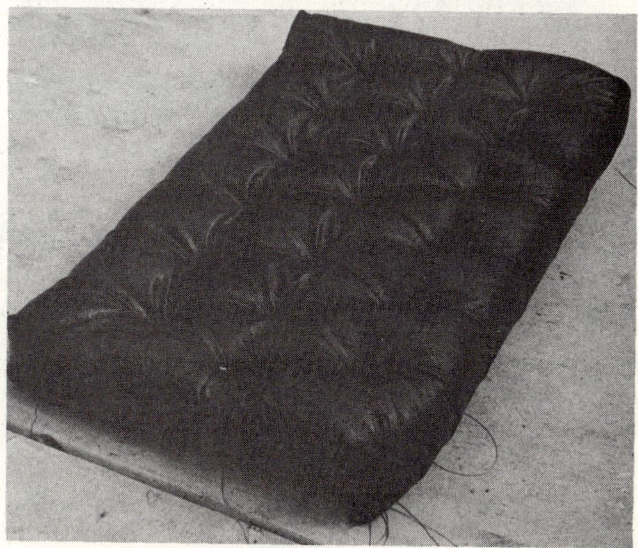

This is a finished seat bottom — button tufted naugahyde over bonded urethane and foam and plywood. This is a low cost, comfortable, attractive, durable item.

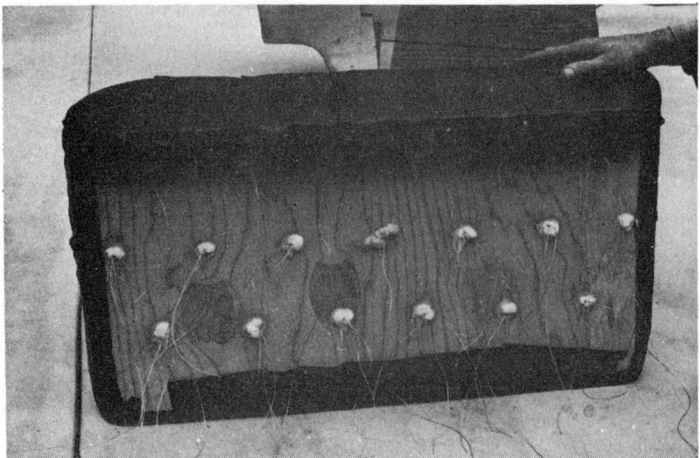

*Underside of completed seat bottom shows the location of all of the buttons and how the material was pulled around the edges of the seat for final gluing and stapling.*

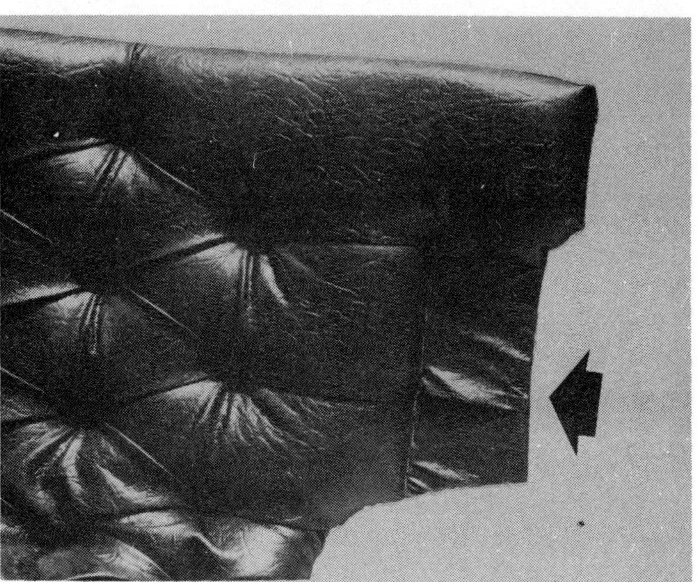

*The recessed portion of this seat former is devoid of buttons or padding — simply material tightly stapled over the former. This area (arrow) tucks down behind the seat bottom.*

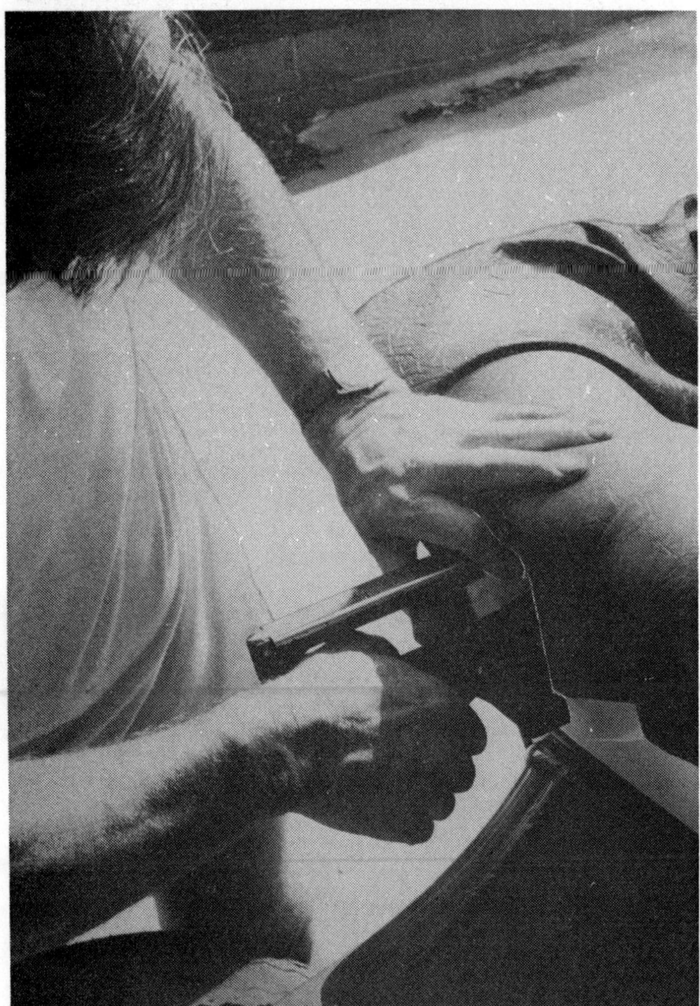

*Once the upholstered seat former is in place in the car and jammed into position, a staple gun is used to attach the upholstery to the underside of the tack strip.*

*The limitations of button tufting lie primarily with the imagination of the builder. This square pattern is quite rich looking with a luxurious roll at the top. Note that the non-tufted roll does require piecing to eliminate creases.*

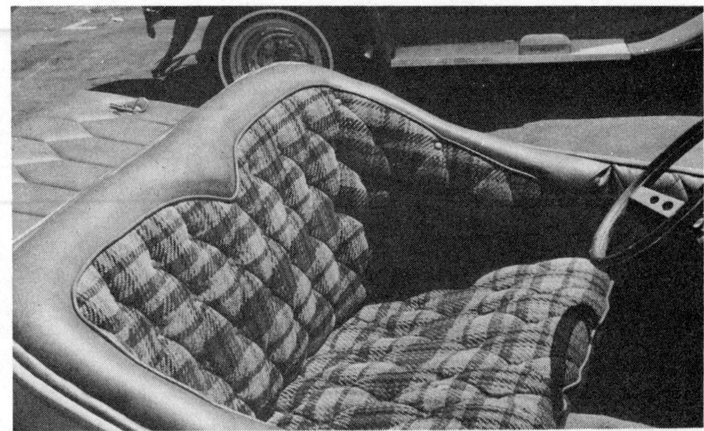

*In this example of tufting, fabric is combined with vinyl material. This is far more difficult than using one material. Note how the diamond pattern is carried out on the shortened pickup bed and the door.*

This is a prime example of combining buttons with stitching. This is quite lush and tastefully done, but unless you can run a professional sewing machine, you'd best pass this version of button tufting.

As you can see, button tufting lends itself well to applications on door panels.

# How To Tie The Button Knot

A very essential part of tufting is in keeping the buttons sunk down at the proper level in the padding. After going to all the work to do your tufting, you sure don't want the tufting to start coming undone the first time you go for a ride. A button knot is the correct knot for the situation and properly tied, it will serve it's purpose in anything you tuft. The button knot allows the button to be tightened — but not loosened.

A button knot is very easy to tie, but you should get in a little practice before getting into a major project. After learning how to tie the knot, play with it and note that one end of the cord is used to tighten the button in the upholstery and the other end of the cord is used to tighten the knot after you get the button where you want it.

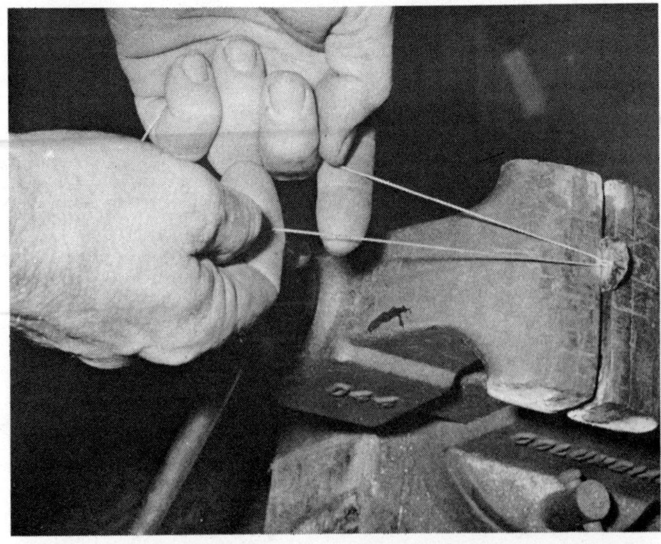

Here's how to practice the button knot. Clamp a button in a vise as shown, run the cord through the eyelet and hold the cord as shown — one end of the cord in the left hand, the other in the right hand. Complicated, huh?

With the left index finger, hook the cord on the right and cross it on top of the cord on the left. When done correctly, this will twist the cord so it crosses twice (note arrows). If you can get this far, you can tie the knot.

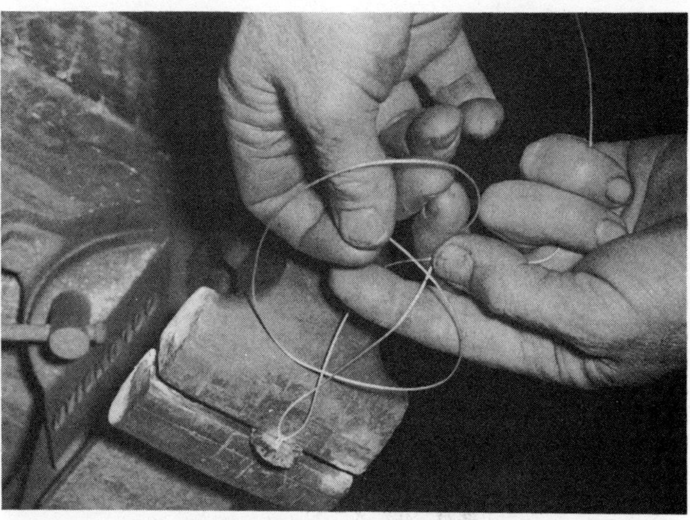

Take the end of the cord now held by the right hand, pass it through the large loop, pass it back through the loop you just created and pull firmly to complete the knot. That's it.

The end of the cord now held by the left hand will now move the knot toward the button. This will sink the button into the material. The other end of the cord tightens only the knot. Play with this until you get it into your headbone how you can move the button down into the upholstery with one end of the cord, and how you can tighten it where you want it with the other end.

# Rod Building Tips III

This tail light support bracket is not stock, neither is the bumper; but both are so tastefully executed they appear to be stock — but better than. This is precisely why you take a second look.

Adding letters to a valve cover like this is a plenty slick and eye-catching idea — even though the engine is a Chevy!

Adding a late model hood scoop on an early car usually doesn't come off very well, but in this case, the scoop looks right at home.

Someone machined up this slick little rain gutter clip which is used to secure the end of a whip antenna when it is not in use. Nice.

This simple, effective roadster dash is easy to make and very low cost. Note that body welting is used about the perimeter of the engine-turned panel.

Commercially-oriented early rod bodies lend themselves well to a bit of sign painting.

This reworked forties vintage dash certainly didn't come stock in this roadster, but is very well executed. Note the gold leafing on the glove compartment script.

(Right) This fine example of rod building shows just how different a Model A outfitted with a '32 radiator shell can be. Horizontal bars in the shell, plus a headlight bar running through the shell plus a very thin hand-fabricated bumper will always set this car apart from the crowd.

This turnbuckle being used as an adjuster on the alternator is a prime example of innovative rod building.

Even though this shortened radiator and shell can't accommodate a full shroud, this is a nice effort to help pull the air through the radiator.

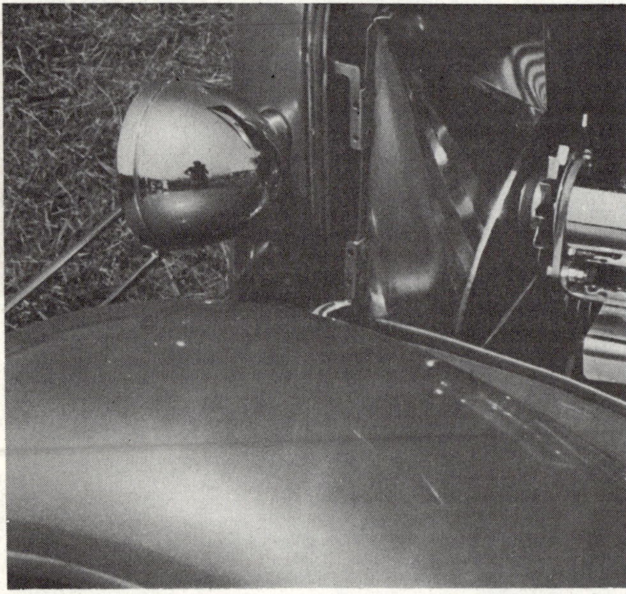

A sheet metal shop can handle a one-of-a-kind shroud such as this — which not only adds some class to the engine compartment but will definitely aid in curing any overheating problems.

This A roadster pickup is very well proportioned. The "grille" for the tailgate is unique.

*There are several ideas here for an early car. The longer you look at it, the better that bumper bracket and bumper look. A heavy gauge rectangular exhaust tip is just unusual enough to draw some attention and that one-of-a-kind exposed tailgate rounds out the picture.*

*Here's another slick trick for latching a pickup tailgate. A chromed lug nut is jammed against a latching plate supported by a chromed stud coming out of the bed side support.*

*Here's another approach to a fan shroud. This is a simple but effective design that can be taken care of by most any sheetmetal shop.*

*This spring hanger on a suicide front end is a dramatic and natural place for a little artwork. Note also how the hanger has been carefully blended into the front crossmember.*

*The flexibility of wood and the versatility and availability of hand-held routers allow the construction of instrument panels such as this.*

Here's one slick way of hiding the snaps on a pickup tarp. A board is placed just inside the tailgate from side to side of the bed. Note the placement of the snaps on the inside of the board.

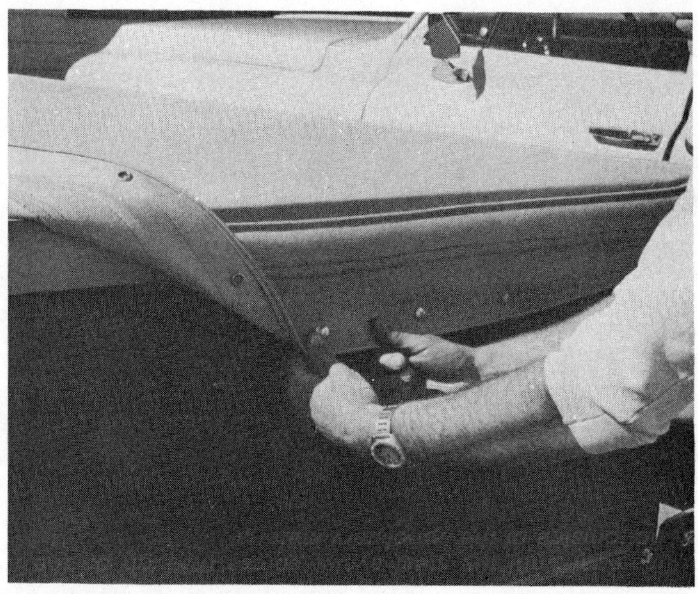

To snap the backside of the tarp, it is pulled firmly over the board . . .

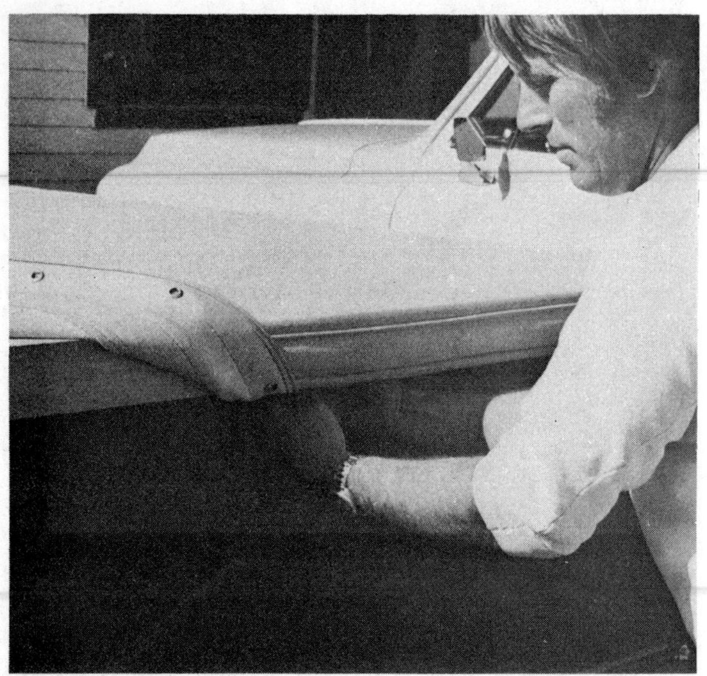

. . . and around it for snapping.

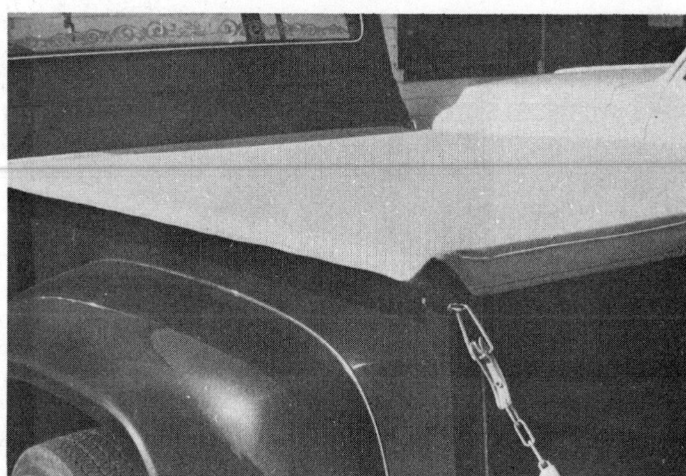

The result is this very clean, "snapless" tarp. This same trick can be used on most any model pickup.

# How To Do Engine Turned Gold Leafing
## A MIND-BLOWING SPECIAL EFFECT FOR ROD, CUSTOM, TRUCK OR VAN

In one form or another, gold leafing has been around for a very long time. Egyptians and Aztecs are just two cultures known to have been acquainted with gilding (or gilting) all sorts of art objects with gold. The purpose, of course, was to make the object more attractive and more valuable via the addition of a thin layer of gold. Gilding has continued through modern times until most of us became familiar with gold leafing (a sophisticated form of gilding) through the gold lettering on the bank window or door.

Simply stated, gold leafing is the applying of real gold in sheet form to some object. It can be in the form of lettering on the lawyer's door, the numbers on a race car, or the belt line trim on a rod. Just keep in mind that we are talking about real gold — gold leafing used by sign painters is normally 23 carat! Without getting into gold paint — which is not leafing — there are three kinds of gold leafing — varigated or multi-colored, plain and engine-turned. If you want to be a nit picker, there are really only two kinds of gold leafing because the plain gold leaf is transformed into engine-turned leafing or left plain by whomever applies it.

Depending on what you intend to gold leaf, this art form leans heavily on patience and a lot of time — which is why gold leafing is relatively expensive to have done, yet quite inexpensive to do yourself. We are indebted to Bob Spina who demonstrated the fine art of applying engine-turned gold leafing along with passing along the tips needed to make your first job go smoothly.

In the first place, you should know that gold leafing and related supplies can be purchased in most art supply stores, sign painters' supply shops and hobby and craft shops. In order to do a small job — such as the one illustrated — you'll need a packet of 25 sheets of 23 carat patent gold leaf manufactured by Gusto Manetti. These sheets are 3⅜-inch by 3⅜-inch by "oh too thin" in thickness. On the back of the package, look for the words "For gilding in the wind" and "Made in Italy." This will set you back about $30.

You should also know there is now imitation gold leafing which is considerably less expensive than the real thing and is extremely hard to detect as being imitation. It can be engine-turned just like the real thing — although it is somewhat more fragile when being turned. A small packet of the imitation leaf (available in varigated or plain) will cost about $3. At this point it is simply a matter of "pays your money, takes your choice." Regardless of the choice you make, you'll need a small can of gold leaf sizing and a small brush with which to apply the sizing. Other material needed will be a small scrap of plush velvet (not crushed), some cotton, a small amount of wax and grease remover, a small amount of clear enamel and a large amount of time and patience.

When preparing to gold leaf, the first step is to thoroughly clean the area to be leafed with the wax and grease remover. Wipe it down with a small clean rag, then wipe it off with a clean, dry rag and then allow the surface to air dry. The next step is to brush on the gold leaf sizing with a small brush. If you are applying the gold leaf to the raised letters on the tailgate of a pickup, the raised letters themselves will serve as a guide as to where the sizing will be applied. If you are applying the sizing (and subsequently the gold leaf) to a flat surface and some guide is needed to contain the sizing, there are two methods which can be used. If the area to be leafed is fairly simple as far as design is concerned, then cellophane or common mending tape can be used to outline the area to be sized. Masking tape is almost impossible to use for this purpose because the sticky side is porous which allows the sizing to seep under it. This also allows the gold leafing to stick to areas you did not want gold — and generally speaking, you have a mess on your hands.

The other method of outlining the area to be sized is to lay out the pattern on butcher paper, run over the perimeter of the design with a pounce wheel, apply the paper to the surface to be leafed, transfer the design and then — using this as a guide — paint the sizing where it belongs. Fortunately, sizing is water soluable as long as it has not set up — so if you goof — the sizing can be wiped off with a damp rag. Such is not the case once the sizing is dry.

It is absolutely imperative that there must be sizing over the entire surface to be leafed. Leafing will not stick where there is no sizing. Keep that in mind before trying to hurry the job. Sizing is almost perfectly clear — and thus difficult to see once it has been brushed on. This can be remedied by adding just a drop or two of paint to the sizing in order to increase visibility. If you are leafing a dark surface, add a drop or two of white or yellow paint to the sizing. If you are leafing a light surface, put a drop or two of black or brown paint into the sizing for contrast.

Don't hurry the job of brushing on the sizing. It must be dry before applying the gold leaf. This is most important. The sizing cannot even be tacky when the leafing is applied. If the sizing is not dry, disaster will surely result. The gold leafing is so very thin it cannot easily be picked up or even held without being on the backing paper it is packaged with. If the leafing is applied to the sizing while the latter is still tacky, the leafing will tear, wrinkle and generally make you want to give up the project in disgust. Normally, if the temperature in the working area is warm and dry, the sizing can dry in one to two hours, but if the air is cool or damp, the drying time can go up to four hours. There is now "five minute sizing" on the market which "goes off" much quicker than standard sizing. If you can find this type of sizing, use it because it will shorten the time you have to spend to complete the job. Test the sizing by gingerly touching it. If it feels to be the slightest bit tacky — back off. If it appears to be dry, attempt to slide your finger across the sized surface. If this produces a squeaking noise, the leafing can then be applied; if it doesn't squeak — back off and wait.

To apply the gold leaf, open the package carefully and flip back a sheet of the leaf and the backing paper that goes with it. Carefully pat the gold onto the sized surface by patting the backing paper toward it. Do not attempt to remove the gold leaf from the backing paper. When the gold sticks to the sized surface, it will automatically separate from the backing paper. If the surface to be leafed is larger than a sheet of leafing, simply use another sheet of gold to cover the area — overlapping the previous sheet by a slight amount.

The leaf will stick to the sizing but not to the other leaf or any of the surface not sized. Continue to pat the leaf in place until the entire surface to be leafed is covered. Pat small scraps of leaf to any uncovered areas until you are satisfied there are no gaps in the gold.

With all of the leafing now in place, but still rough around the edges, gently stroke the leafing with a small ball of cotton. The primary purpose of this operation is to polish the gold, but it will also serve to remove flecks of loose gold and to produce a reasonably clean edge line. Polish the gold until it has a soft sheen to it.

If you just want plain gold leafing, you now have it and a protective coat of clear enamel can be applied. However, for the engine-turned effect, a further step is in order. Place a small ball of cotton inside a scrap of velvet and then twist the velvet around the ball with the plush side of the velvet out. Flatten the ball slightly before pushing it onto the gold and twisting. Do not move the ball sideways or up and down on the surface — just press the ball onto the gold and twist. Continue this process of twisting the velvet ball on the gold until the entire surface is "engineturned."

Naturally, there can be variations to this basic technique. A small wad of cotton can be placed on the eraser end of a pencil and then covered with a scrap of velvet for smaller circles. Also , a very formal pattern of circles can be laid out in contrast to the freehand approach. In either case, the circles should overlap slightly.

At this point the leafing must now be protected. As mentioned earlier, this entails an evenly brushed-on coat of clear enamel. After allowing this protective coating to dry for at least thirty minutes, the gold leafing can then be outlined with pinstriping. In order to hide the edge of the gold leafing, half of the stripe will go on the gold and half on the background surface. It is important to note that the clear must be applied over the gold before the striping goes on — if not, a mistake in the striping cannot be corrected if it is on the gold. After allowing the pinstriping and the first coat of clear to dry overnight, brush on a second coat of clear to protect the gold.

A couple of closing points are in order. Varigated gold leafing should be applied when the sizing is tacky. Plain gold leafing (to be turned into engine-turned leafing) must be applied when the sizing is "squeaky" dry. By keeping the gold leaf surface out of automatic car washes and away from rubbing compounds and idiots who park be sound, engine-turned gold leafing can last for several years on a car driven daily on the streets — much longer if the car is always garaged or covered and is lavished with tender loving care.

Gold leafing should be no mystery to any rod builder. To do the job right takes a minimum of material and a maximum of patience.

*This is an excellent example of how gold leafing can be used to accent a vehicle. The treatment seems to be equally at home on most any vehicle.*

*Real gold leafing is becoming increasingly expensive and difficult to find. The real stuff costs ten to twenty times as much as imitation gold leaf — which works equally as well in most cases, and is available in most art supply and hobby shops.*

*Quick dry synthetic sizing is the bonding agent that attaches the gold leaf to the surface. It is inexpensive and readily available in paint and art supply stores.*

*Paint the sizing onto the area to be leafed. If you miss a spot, the leaf will not stick. if you put the sizing where you do not want gold, you will most likely have gold anyway — so be careful. The sizing is practically clear, so when sizing is applied to a light surface, add several drops of darker paint to the size. This helps you see where the sizing has been applied.*

*It is imperative that the surface to be leafed be clean and free of all wax and grease. Wipe the entire surface down with wax and grease remover several times before the leafing is applied.*

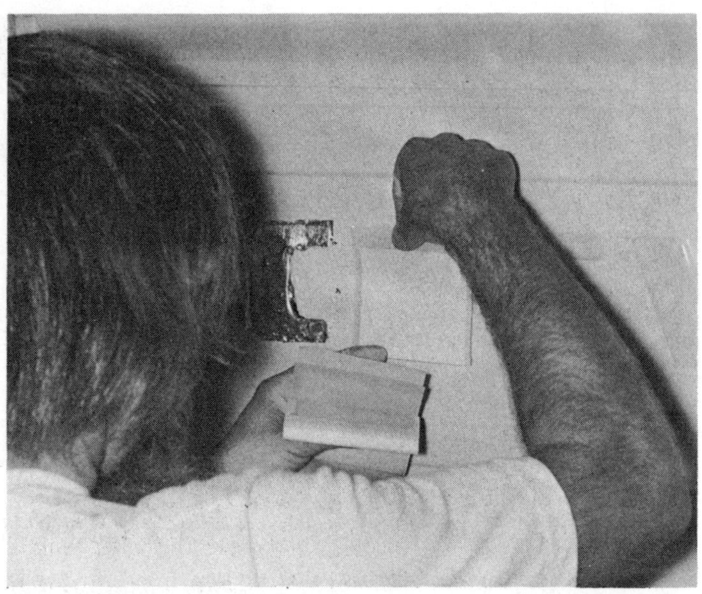

With real gold leaf the sizing must be dry — not tacky when the leaf is applied. With imitation leafing, the sizing must be tacky at the time of application. Apply backing paper and leaf to the surface, then remove the backing paper. If the leafing tears or must be patched for any reason, simply lay more leafing on. Don't worry about overlapping.

To eliminate all of the excess leafing, use a small ball of cotton to simply wipe down the entire area. All of the loose gold leafing will just fall away.

(Right) Firmly push the velvet-covered cotton onto the leafing and twist one-half to three-quarters of a turn. Overlap the previous twist and always twist in the same direction.

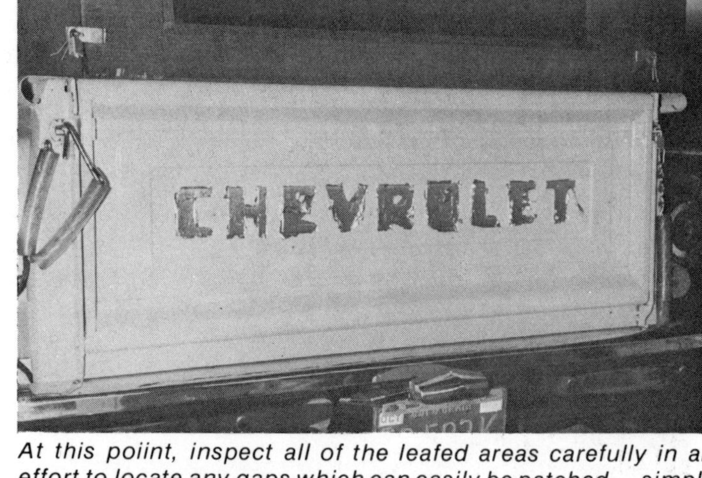

At this poiint, inspect all of the leafed areas carefully in an effort to locate any gaps which can easily be patched — simply by applying more gold leaf.

The first step in engine turning gold leafing is to cover a small ball of cotton with a piece of velvet. It is the velvet surface that "turns" the leafing.

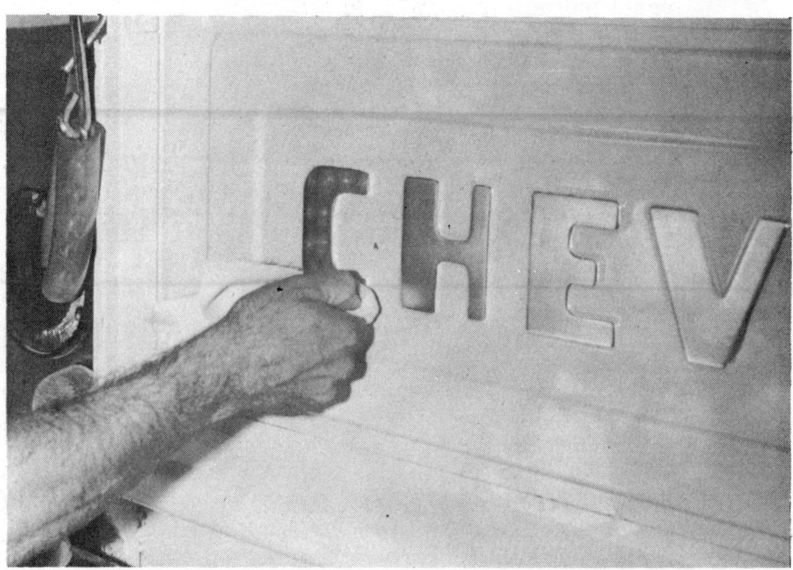

The contrast between unturned and turned leafing is dramatic.

At this point the engine turning is complete, but the job is still unfinished.

Striping around the edges of the leafing gives the job a finished look. After the striping is completed and has dried, the leafing and the striping need to be covered with at least two coats of clear enamel.

Because the gold leafing can serve as such a dramatic highlight, it lends itself well to a shadowing effect such as that shown here. Note also that the gold script serves as a focal point on the very large, flat expanse of tailgate.

Here's a heavy application of varigated gold leafing — a handsome accent to a well-detailed rod.

If you have the patience of a prophet, gold leafing can be applied as a stripe — fire truck style.

If you can't letter like this, but can gold leaf work something out with the local sign painter who can lay out the letters for you to finish. This is a very nice piece of work on an early panel truck.

Here's another case of engine-turned gold leafing being applied as a wide stripe. The contrast to the black paint on this early Ford is dramatic.

Obviously the sky is the limit when it comes to creating accents with gold leafing. Designs such as this are easy to work out on paper so they can later be transferred to the sheet metal for the effect shown here.

# The Basics of Striping

Nothing is quite the finishing touch for a rod, custom, truck or van like genuine pinstriping. The stripes can be mild or wild — byt if they are in good taste and well done, pinstriping is a plus for most any vehicle you can name.

Nothing else is quite as likely to strike fear in the heart of a rod builder than the suggestion that he do his own striping. Some of the fear is in jest and some of it is well justified (there are some of us that just can't stripe). But, we've known any number of artistically inclined rod builders over the years that were intrigued with the idea of doing some striping, but simply did not know how to get started. Hopefully, the words and photos contained here will help.

To get the straight skinny on striping, we had a session with a most talented guy by the name of Bob Spina. Bob has done it all when it comes to automotive painting — candy, pearl, fading, fogging, paneling, lace, murals, gold leafing — you name it. But it is striping that is Bob's first love when it comes to accenting a car. He started as a kid — and has been laying a mean line ever since.

"I remember we lived in this old house when I got started. I striped everything is sight for practice. Striping window glass is neat because it is so easy to clean off and it's also very easy to see your mistakes. I even striped the boards on the floor."

That statement from Bob tells you a lot about the kind of many and talent he is, and it should tell you even more about what it takes to make a good striper — practice, and plenty of it. If you truly want to "lay a line," then you will want to acquire a small supply of brushes, paints and various other bits of inexpensive trivia and start practicing. As you now know, the practicing can be done on most any solid object — and since striping is easy to remove while it is still wet you can stripe the same chair, refrigerator or filing cabinet over and over again.

For the most part, the suppies needed for striping are very inexpensive and readily available. To do the job right, it is essential that any striper have the right equipment — starting with the correct brushes. Striping brushes are just that; they were not intended to be used for anything else. Trying to stripe without a striping brush is like doing you-know-what against the wind. Striping brushes come in a variety of sizes and the sizes are designated on the handle of the brush by a number — such as 00, 0, 1 or 2. The 00 or "double aught" is the smallest of this group and the brushes get larger from there on up. The four brush numbers just noted will handle just about any striping that comes along and a good set of brushes will last indefinitely if cared for; so you can see that at about two bucks per brush that your investment is minimal.

Striping brushes are also called daggers or sword

stripers and can be purchased in art supply stores, sign supply stores and some of the larger office supply stores. One of the highest quality brands of striping brush on the market is Grumbacher. This brand is imported and sometimes quite difficult to find, but any good striper will tell you that the quality is worth the search.

Striping can be done with enamel-based paint or with a special brushing lacquer, but the beginner would be well advised to stick with a sign painter's paint known as Bulletin made by Sherwin-Williams. It has a good flow and provides good coverage. "1-Shot" is another quality brand of enamel widely used for striping. Both are available in a wide variety of colors and, of course, they can be mixed to provide an infinite array of hues. Both brands of sign painter's enamel mentioned flow very smoothly and have a relatively fast set up time of five or six hours, but can easily be removed with a turpentine soaked rag within a couple of hours of application.

Thinner should always be available to achieve the right consistency of paint. For oil-based paint, turpentine is best, and lacquer thinner should be used for lacquer paints. The lacquer paint should be used only to stripe a lacquer-painted car since lacquer eats away enamel paint. On the other hand, oil base is OK for all types of paint . . . so when in doubt, use oil base.

Whatever you are striping must be clean. Be sure to wipe the surface to be striped with a wax and grease remover. Saturate a clean rag with the solution and wipe the surface several times. Don't confuse thinner with wax and grease remover — since thinner will not do the job of getting rid of wax and grease. After the wax and grease remover has been wiped off with a clean, dry rag, the surface is ready to be striped.

The supplies needed for striping are quite simple and straightforward. A small container of thinner (a baby food jar works fine) is needed to clean the brush before each loading. In addition to the three or four brushes, the thinner and the paint, you should have several clean shop rags and some clean paper, cardboard or old magazines on which to wipe the brushes while loading them with paint. Spina — along with a number of other stripers — prefers old car magazines, primarily because there are always stacks of them around. This brings up a point which is well worth taking into consideration — an abundance of car magazines is a good source for fresh ideas. Some simple line or technique used on a van paint job just might spin off an idea for a treatment to a race car or sports car, and so on.

While you are wiping down the surface to be striped with the wax and grease remover, have the brush to be used soaking in the small container of thinner or turpentine. This removes the oil from the bristles (more about this later). One of the biggest factors in striping is in preparing the paint. The paint must flow, but it also must have a certain amount of drag to give the brush somewhat of a rudder action. Next to the magazine or paper, set the

can of paint and the container of thinner. Dip the brush into the paint and then into the thinner. On the paper, work the brush back and forth on both sides, evenly distributing the paint into the brush. This is called "loading the brush." This must be done every time a line is started. When loading a brush, lay the brush practically parallel with the pages and work the bristles along in a squiggly motion toward you. When pulled from the paper, the properly loaded brush will not drip, but will droop slightly under the weight of the paint. You'll also note that the brush will assume the shape of a dagger when properly loaded. When you have this right, the flow of paint from the brush will be even and smooth.

Take a look at the photos and you'll see that the brush is always held between the thumb and the forefinger and that the forefinger is always held up or away from the rest of the fingers which are laid upon the surface being striped as a guide. At this time it should be pointed out that if the brush happens to have a square handle, it should be filed and sanded down until it is fairly round. This is because when making a curve, the brush must be rotated into the curve during the stroke — and it is most difficult to rotate a square brush handle between thumb and forefinger while trying to maintain a stripe!

Glass is an excellent surface to practice on when learning to stripe. It is flat, smooth and can easily be cleaned and used again. Get a piece of glass (even framed, removable windows) that can be laid flat to begin your striping practice. Later as you get better and gain confidence, the glass can be tipped up to simulate the side of a car. Start practicing straight lines. Find a width of line you feel comfortable with and practice it until you begin to feel comfortable with the brush and paint. As you progress, practice making parallel lines of the same thickness across the surface. Then try thinner and thicker lines. Don't shy away from practicing thin lines, because you'll soon discover that it is more difficult to maintain a constant width thick line over a long pull that it is to maintain a thin one.

When learning to stripe, try to maintain a brush angle of about 45 degrees to the surface. Just slowly lower the tip of the brush to the surface and begin to slowly pull the brush toward you. If you pull the brush too fast, the line will be too thin (in terms of paint content), and if you pull too slowly the chances are that the line will be fat and shakey. On very long lines, the brush will need to be reloaded. This takes some additional practice, but the technique is to simply pull the brush up, reload and pick up the line again, but go over the last five or six inches of the existing line so the hand will be steady again and the line will continue to be straight.

While all of this is going on and you are practicing lines, be aware of the condition of the paint. If it begins to noticeably thicken on the brush while pulling a line, dip the brush quickly into the thinner and then "massage" the paper again as when you originally loaded the brush.

Until you get very good at striping, you'll probably want to rely on a guide to help you pull a straight line — a piece of chrome, a molding, a break in a surface or an edge. Masking tape or a magnetic strip can also be used. Pleace any of the fingers not used to hold the brush on the guide and let this serve to keep the tip of the brush in alignment with the guide. Chances are that when you first start out that you will be shakey and make a lot of mistakes. Don't let it bother you since the line can easily be wiped off with the turpentine on a rag. One fo the first things you'll have to concentrate on after getting a feel for loading the brush is to move the entire hand and arm instead of just the fingers holding the brush.

Get a good feel for what it takes to pull a straight line before moving on to the more difficult stuff. One of the most difficult things to get a grasp of is how to go around a corner or turn with the striping brush and keep the line width constant and smooth. This takes a lot of practice and involves rolling the brush between the forefinger and thumb as you stripe into the turn. If this is not done, the bristles on the outside of the turn will tend to pull away from the brush and leave small spurs or "whiskers" on the outside of the line. Practice drawing circles with the striping brush — try both clockwise and counterclockwise and stick with the method that is easiest for you. Because the hairs of the brush do want to spring outward while striping a curve, you'll want the consistency of the paint a little heavier that you normally would use. Once you have curves and circles mastered, work on striping the letter "S." Once you have that down, you should have full confidence to strip anything from a model airplain to a locomotive.

Because so very little is involved in pinstriping in the way of equipment, it is very easy to maintain. Good striping brushes improve with age. They "load" better and they are simply easier to use. When a striping job is finished, wash the brush out with turpentine and then saturate it with regular engine oil. Form the bristles of the brush into the dagger shape and then lay the brush on a clean surface and stroke the bristles until the upper edge is straight and the lower edge curves up to form the dagger shape. Store the oil-saturated brush in a place where it will remain saturated and will not be disturbed.

To be reasonably good at striping, you must be somewhat artistically inclined and you must be able to hold a brush steady. If you are severely lacking in either area, striping is not for you. But for less than ten bucks you can sure find out if you can acquire the skill. We've known and watched a lot of talented stripers — and to a man they all stress practicing until you have the confidence and skill to make the brush do what you want it to do. We've been in the homes and garages of stripers and found solid evidence that they do their share of practicing — on mail boxes, tool boxes, garden tool handles, wheelbarrows, garbage can lids, furniture and even woodwork inside the house. Start practicing.

*This limited amount of equipment — plus some skill and a lot of patience and practice — is about all you need to "lay a line." Sign painter's enamel, thinner, brushes and an old magazine for loading a brush is all that's needed to do a professional striping job. A quart of turpentine will do just fine for cleaning brushes before and after striping and also for actually thinning the enamel. Cost for all of this is next to nothing.*

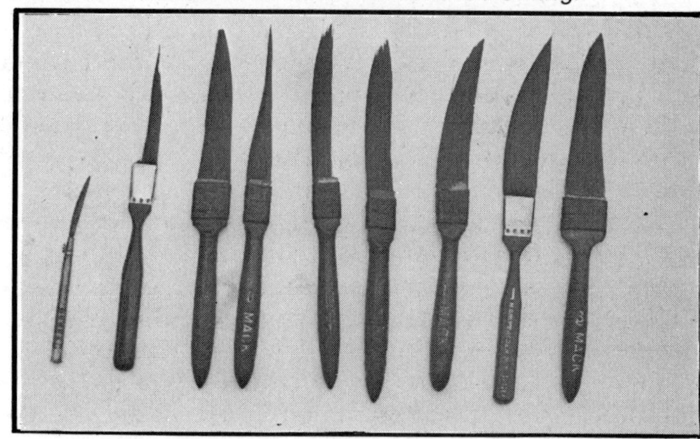

*These striping brushes show the variety of sizes and shapes of daggers available. All of these are Grumbacher or Mack. Four "part numbers" — 00, 0, 1, and 2 will handle most any striping job you can dream up.*

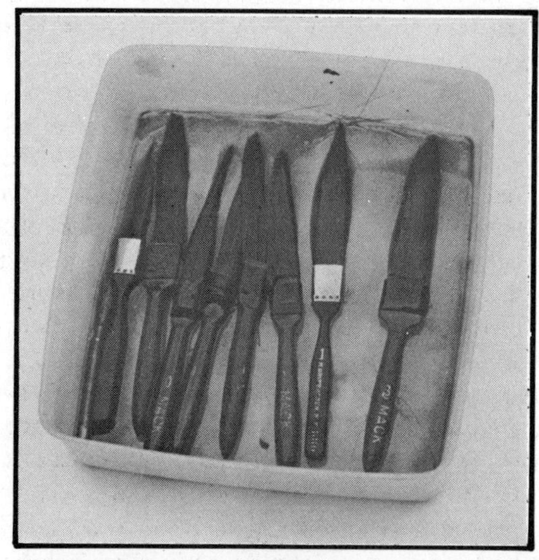

*When not in use, striping brushes should be stored in oil. Ordinary motor oil will do just fine. Never store a striping brush with the weight of the brush standing on the bristles.*

This is the technique to use for loading a striping brush with paint. Dip it in the paint and work it back and forth across a page of an old magazine. When the bristles droop under the load of paint — with no dripping — you are ready to stripe.

A visit to your local magnetic sign shop will uncover these magnetic strips which are great guides to beginning stripers. They are very flexible, low in cost and can be used over and over again.

This is the sort of situation that can baffle a beginning striper and yet the situation almost solves itself when the magnetic tape is used as a guide.

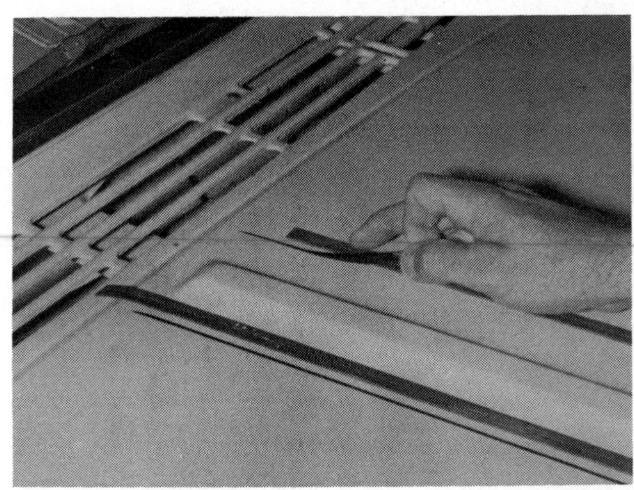

With the magnetic guide in pace, the beginning striper can concentrate on pulling a line which is evenly weighted — neither thick nor thin — while the guiding finger on the magnetic strip makes sure the line is straight. Note here how the brush is held between forefinger and thumb and that it lays down very close to the panel.

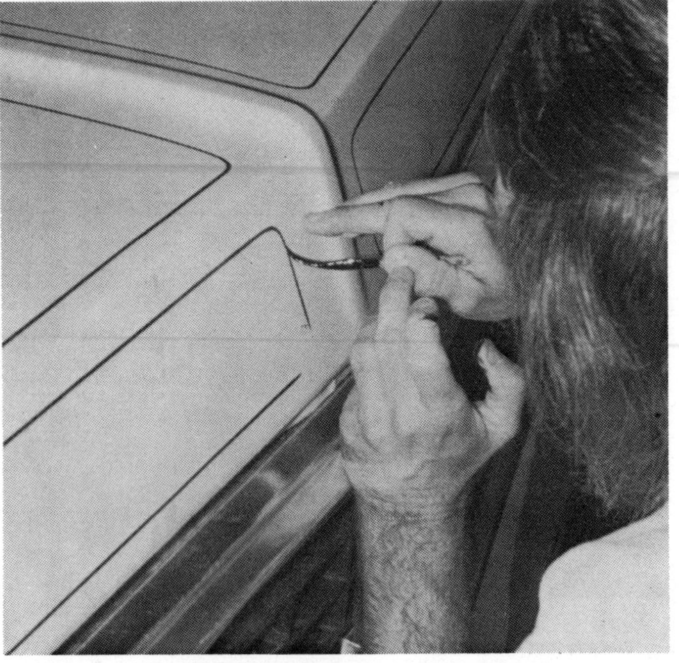

Keep in mind that you don't stripe an entire car without ever lifting the brush. The right side of this hood has been completed, while the left side shows how all of the straight lines were completed before being joined at the corners.

Spina's stance is indicative of the concentration needed for flawless striping. Note that in this instance a break in the sheetmetal is being used for a guide.

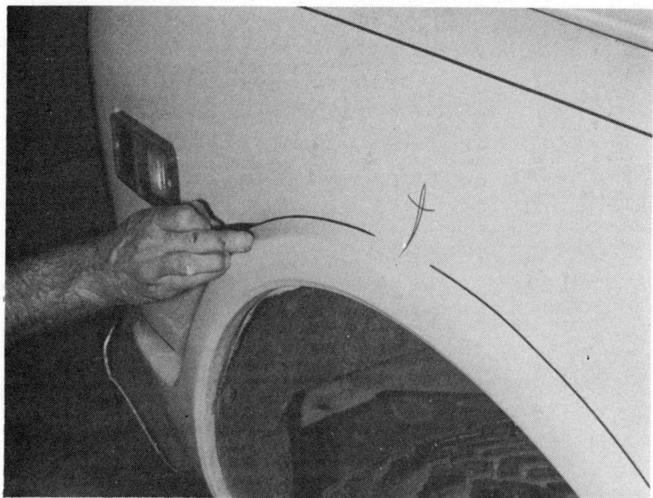

When a number of lines are being used to effect a design, it is important for the striper to have a grasp on which lines go down first. In this case, the small accent — or cross — was applied to the very center of the wheel well before the other lines were laid.

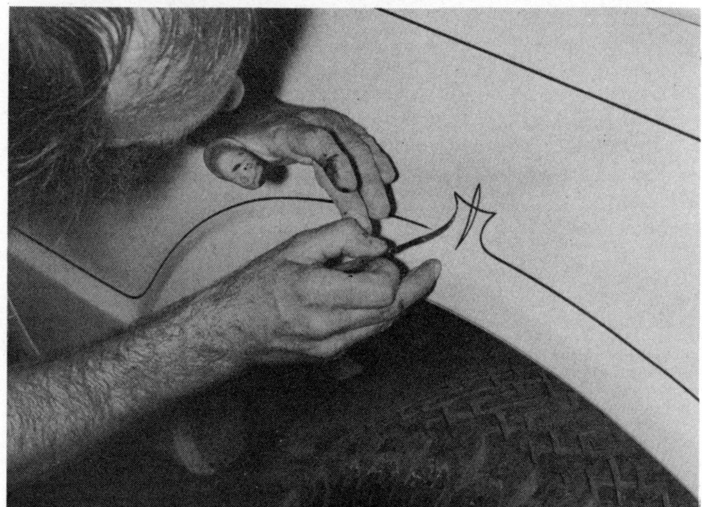

As the various lines are joined, there should be a flow and a reason for the line being there. Note also how the left hand is being used to steady the right hand in this case.

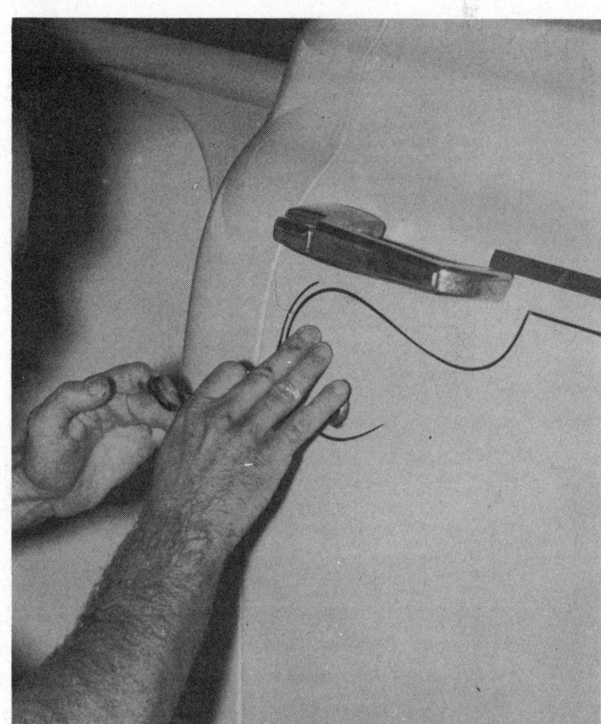

"Curly Qs" take somewhat more practice than a straight line, but the position of the middle finger on the right hand shows how the brush is used to pivot around that middle finger in order to pull a smooth curve.

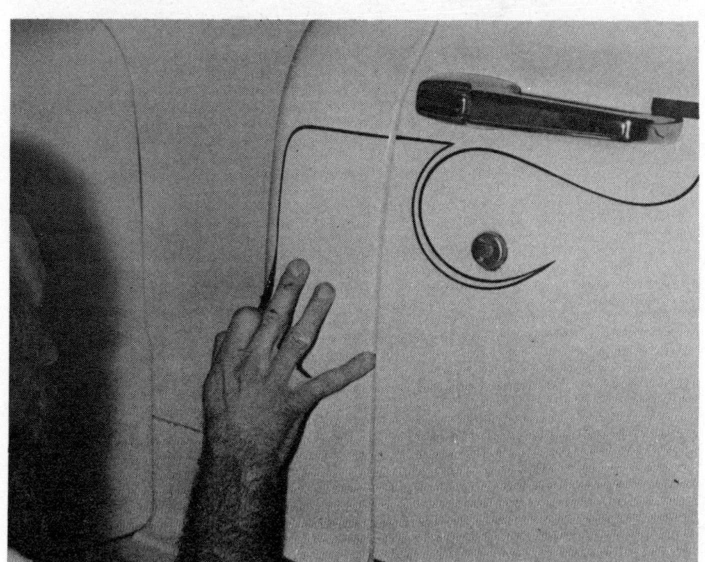

The important thing to note here is how the little finger is being used in the door jamb as a guide for the line down the side of the curved panel.

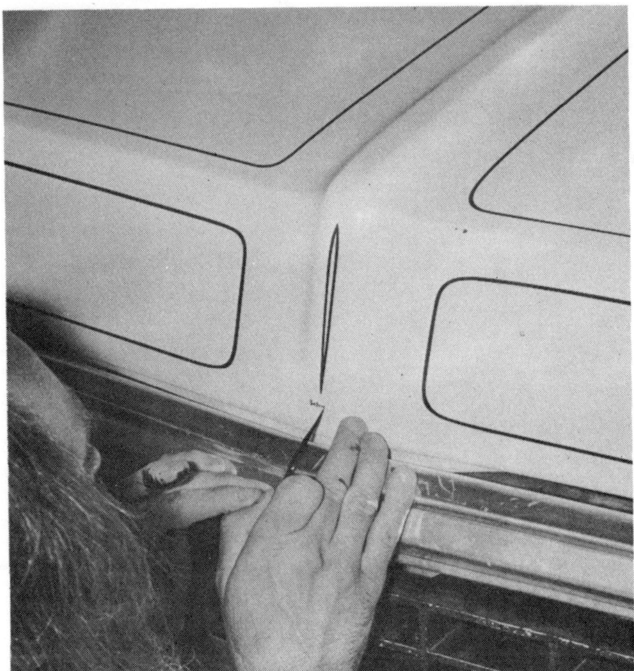

Naturally, the finishing touch to any striping job is the signature. The name should be able to be covered with a dime — regardless of the fact that your name is George Washington Stupenaggle.

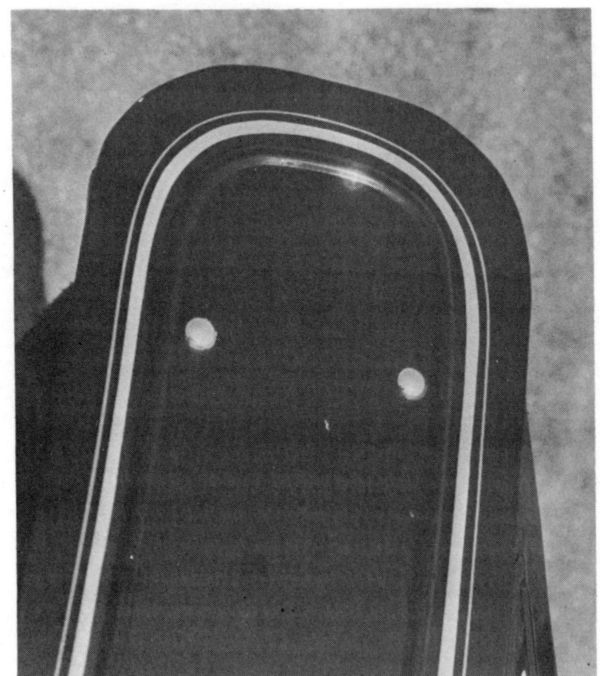

There are all sorts of combinations of thin and wide lines that give various effects on different types of cars. Only experimentation and practice will allow a striper to get full benefit from the various combinations.

Ah yes, multi-colored, free-handed striping for the masses such as this will bring the troops to their knees and will make them all wonder how you ever learned to do all of that.

If lettering is to be used on a rod, make sure the striping complements both the lettering and the vehicle. Such is the case here.

# Rod Building Tips IV

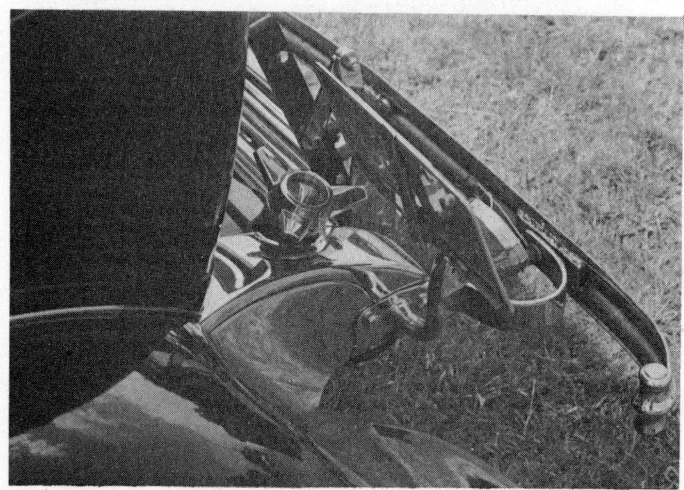

Chroming the rear frame covers on this early car gives a rich custom look as does the wheel cover "spinner" that now serves as a cover for the gas cap.

A half a day in a wrecking yard will turn up any number of neat ideas for a rod or street machine like this lighted vanity mirror. In this case the lights are turned on when the cover is flipped up. Any number of luxury cars feature this accessory.

If your necktie should become caught in the fanbelt while checking timing, injury could result (see arrow).

This A roadster built by John Buttera is a prime example of clean design and excellent workmanship. Note the straightforward approach to everything.

A number of small things contribute just as much to the overall look of a rod as one large item. Note the double beauty ring wheel, the brake line coming through the frame and the shock mount on the frame for starters.

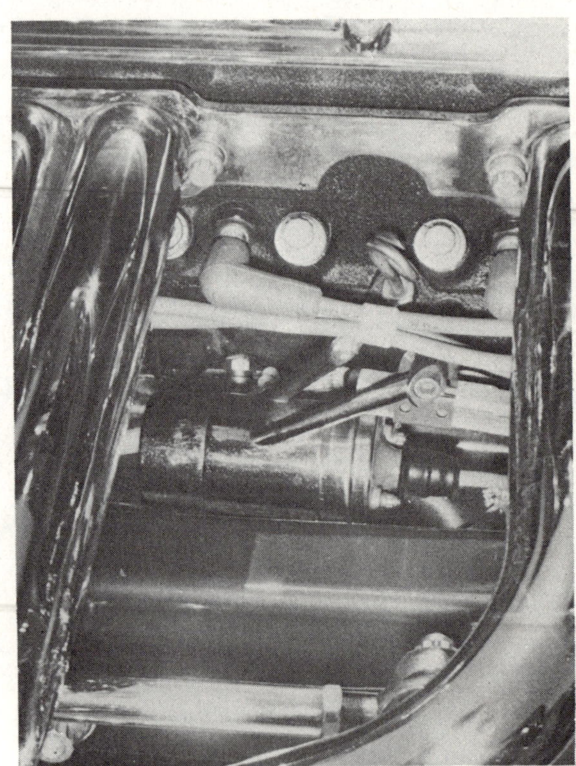

Check the placement and design of the coil bracket on this small block Chevy.

Here's an attractive, low cost, easy-to-build dash and console for an early car. Wood is a good alternative to metal in many applications in rod building — this is especially important to remember if you do not have access to common metal working tools.

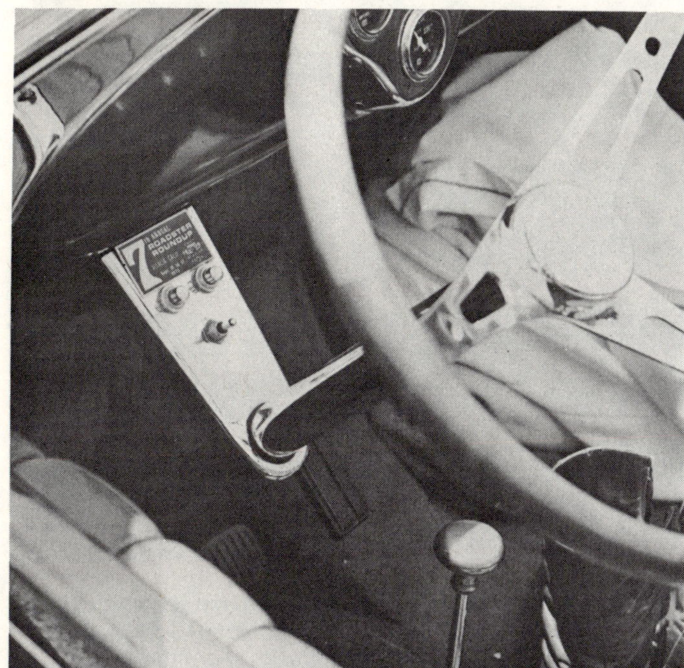

We're seeing an increasing amount of "double function" pieces in contemporary rod building — such as this steering column hanger on a Model A based rod.

This fabricated heat shield on an antique race car just might serve as an inspiration to someone with an early roadster and exposed exhaust.

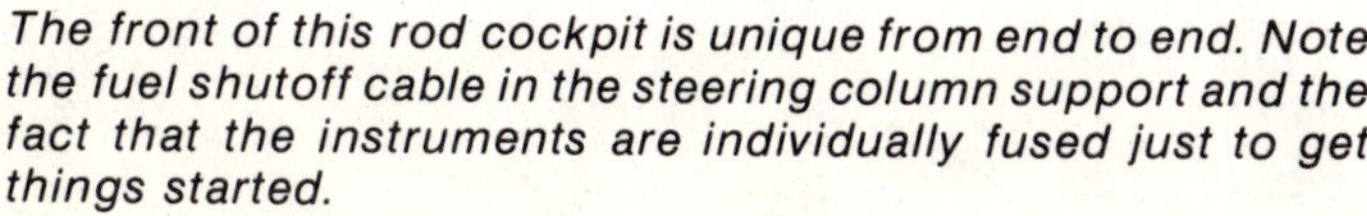

The front of this rod cockpit is unique from end to end. Note the fuel shutoff cable in the steering column support and the fact that the instruments are individually fused just to get things started.

Look close and you'll see that the luggage rack on this T roadster does not fasten to the body — but rather to two removable stanchions on either side of the deck. Thus when the rack is removed, there is no evidence it was ever there.

This dual purpose steering column support not only houses two instruments and an eagle, but directional signal indicator lights as well.

This upholstered steering column support is showing a little wear but the simple lines and the function of containing the ignition switch is "right on."

There is obviously more to this setup than meets the eye. Between the tail lights note the trailer light plug and that step into the camper is obviously designed to hold a trailer ball. Note that the license plate light is offset to one side to minimize damage from foot traffic. Someone was obviously doing some thinking.

If you can get past the glitter and glare of the chrome, check out the path of the spark plug wiring and the use of steel braided hose running into the firewall at a common junction point.

Here's a big block Chevy in a roadster with one of the most unique alternator brackets we've ever seen. Tubing is used for the main member and a piece of slotted steel has been welded to the underside of it.

(Left) A radiator catch can is certainly compatible with early car styling if this much though is given to execution of a good design.

No, this ain't no furrin car — but a one of a kind, home-made rod body. We think the dash represents a neat approach to an early car, and it's something you don't see every day.

Something as simple and subtle as cutting holes in a small panel can give dramatic results on any model car.

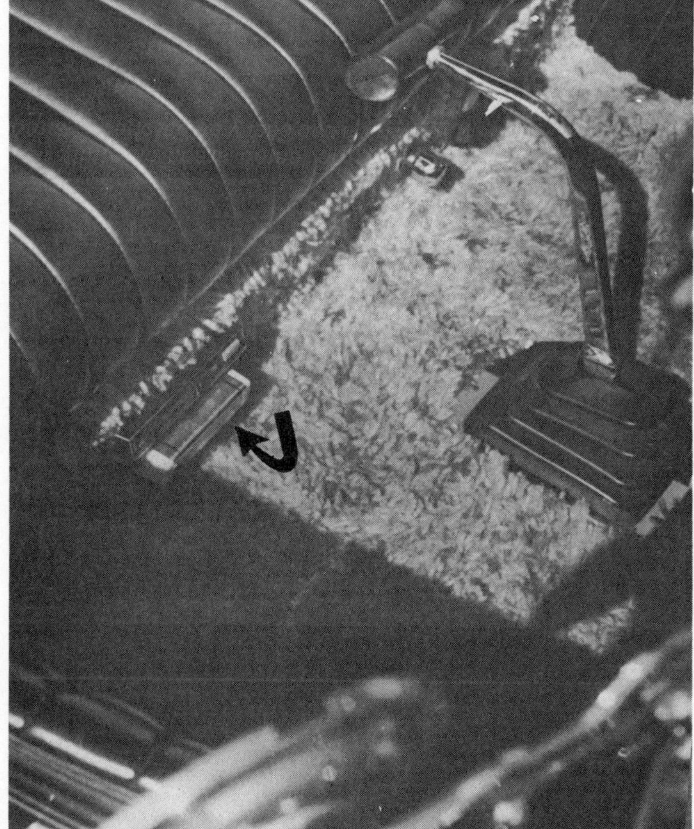

Early roadsters are short on space everywhere you look. Tucking a tape deck under the seat is a slick idea to remember for a lot of rod building projects.

Here's an approach to a door panel you don't see every day. This late model GM pickup sported a full headliner of mohair to complement the six-way power seat in the same material. Nice.